An Introduction
to Surface Analysis
by XPS and AES

An Introduction to Surface Analysis by XPS and AES

John F. Watts
University of Surrey, UK

John Wolstenholme
Thermo VG Scientific, East Grinstead, UK

WILEY

Other Wiley Editorial Offices

John Wiley & Sons, Inc., 605 Third Avenue, New York, NY 10158-0012, USA

Wiley-VCH Verlag GmbH, Pappelallee 3, D-69469 Weinheim, Germany

John Wiley & Sons (Australia) Ltd, 33 Park Road, Milton, Queensland 4064, Australia

John Wiley & Sons (Asia) Pte Ltd, 2 Clementi Loop #02-01, Jin Xing Distripark, Singapore 0512

John Wiley & Sons (Canada) Ltd, 22 Worcester Road, Rexdale, Ontario, M9W 1L1, Canada

Library of Congress Cataloging-in-Publication Data

Watts, John F.
  An introduction to surface analysis by XPS and AES/John F. Watts,
  John Wolstenholme.
     p. cm.
  Includes bibliographical references and index.
  ISBN 0-470-84712-3 (cloth : alk. paper) — ISBN 0-470-84713-1 (pbk. : alk. paper)
    1. Surfaces (Technology)—Analysis.   2. Electron spectroscopy.   I.
  Wolstenholme, John.   II. Title.
  TP156.S95W373 2003
  620′.44 — dc21   2002153114

British Library Cataloguing in Publication Data

A catalogue record for this book is available from the British Library

ISBN 0-470 84712 3 (Hardback)
     0-470 84713 1 (Paperback)

Typeset in 10.5/13pt Sabon by Thomson Press (India) Ltd., Chennai
Printed and bound in Great Britain by TJ International Ltd, Padstow, Cornwall
This book is printed on acid-free paper responsibly manufactured from sustainable forestry,
in which at least two trees are planted for each one used for paper production.

Contents

Preface

When one of us (JFW) wrote an earlier introductory text in electron spectroscopy the aim was to fill a gap in the market of the time (1990) and produce an accessible text for undergraduates, first year postgraduates, and occasional industrial users of XPS and AES. In the intervening years the techniques have advanced in both the area of use and, particularly, in instrument design. In XPS X-ray monochromators are now becoming the norm and imaging has become commonplace. In AES, field emission sources are to be seen on high-performance systems. Against that backdrop it was clear that a new, broader introductory book was required which explored the basic principles and applications of the techniques, along with the emerging innovations in instrument design.

We hope that this book has achieved that aim and will be of use to newcomers to the field, both as a supplement to undergraduate and masters level lectures, and as a stand-alone volume for private study. The reader should obtain a good working knowledge of the two techniques (although not, of course, of the operation of the spectrometers themselves) in order to be able to hold a meaningful dialogue with the provider of an XPS or AES service at, for example, a corporate research laboratory or service organization.

Further information on all the topics can be found in the Bibliography and the titles of papers and so on have been included along with the more usual citations to guide such reading. The internet provides a valuable resource for those seeking guidance on XPS and AES and rather than attempt to be inclusive in our listing of such sites we merely refer readers to the UKSAF site (www.uksaf.org) and its myriad of links. Finally, we have both been somewhat perturbed by the degree of confusion and sometimes contradictory definitions regarding some of the terms used in electron spectroscopy. In an attempt to clarify the situation we have included a Glossary of the more common terms. This has been taken from ISO 18115 and we thank ISO for permission to reproduce this from their original document.

John F Watts
John Wolstenholme
Guildford Surrey UK
East Grinstead West Sussex UK

Acknowledgements

There are many people who have influenced the development of this book: students, research workers, customers and potential customers, and many colleagues too numerous to mention, both at the University of Surrey and Thermo VG Scientific. At the University of Surrey the staff and students associated with The Surface Analysis Laboratory have provided a stimulating and exciting atmosphere in which to work. Professor Jim Castle has been an inspiration not only to the authors (and one in particular!), but to the entire applied electron spectroscopy community. We both wish him well in his retirement. In addition, Andy Brown and Steve Greaves must be thanked for the production of many of the spectra and other graphics used in the text. At Thermo VG Scientific, Kevin Robinson and Bryan Barnard have provided stimulating leadership in their respective fields, and have provided invaluable assistance in certain areas of the text. Present and former members of Thermo VG Scientific's Applications Laboratory are gratefully acknowledged for their assistance in providing data and valuable information for inclusion in this volume.

Certain figures and data have been reproduced from other sources and we thank the copyright holders for their permission to do so. The cover design makes use of original computer graphics generated by Paul Belcher (Thermo VG Scientific).

1 Electron Spectroscopy: Some Basic Concepts

1.1 Analysis of Surfaces

All solid materials interact with their surroundings through their surfaces. The physical and chemical composition of these surfaces determines the nature of the interactions. Their surface chemistry will influence such factors as corrosion rates, catalytic activity, adhesive properties, wettability, contact potential, and failure mechanisms. Surfaces, therefore, influence many crucially important properties of the solid.

Despite the undoubted importance of surfaces, only a very small proportion of the atoms of most solids are found at the surface. Consider, for example, a 1 cm cube of a typical transition metal (e.g., nickel). The cube contains about 10^{23} atoms of which about 10^{16} are at the surface. The proportion of surface atoms is therefore approximately 1 in 10^7 or 100 ppb. If we want to detect impurities at the nickel surface at a concentration of 1 per cent then we need to detect materials at a concentration level of 1 ppb within the cube. The exact proportion of atoms at the surface will depend upon the shape and surface roughness of the material as well as its composition. The above figures simply illustrate that a successful technique for analysing surfaces must have at least two characteristics.

1. It must be extremely sensitive.

2. It must be efficient at filtering out signal from the vast majority of the atoms present in the sample.

This book is largely concerned with X-ray photoelectron spectroscopy (XPS) and Auger electron spectroscopy (AES). As will be shown, both of these techniques have the required characteristics but, in addition, they can answer other important questions.

1. Which elements are present at the surface?

2. What chemical states of these elements are present?

3. How much of each chemical state of each element is present?

4. What is the spatial distribution of the materials in three dimensions?

5. If material is present as a thin film at the surface,

   (a) how thick is the film?

   (b) how uniform is the thickness?

   (c) how uniform is the chemical composition of the film?

In electron spectroscopy we are concerned with the emission and energy analysis of low-energy electrons (generally in the range 20–2000 eV[1]). These electrons are liberated from the specimen being examined as a result of the photoemission process (in XPS) or the radiationless de-excitation of an ionized atom by the Auger emission process in AES and scanning Auger microscopy (SAM).

In the simplest terms, an electron spectrometer consists of the sample under investigation, a source of primary radiation, and an electron energy analyser all contained within a vacuum chamber preferably operating in the ultra-high vacuum (UHV) regime. In practice, there will often be a secondary UHV chamber fitted with various sample preparation facilities and perhaps ancillary analytical facilities. A data system will be used for data acquisition and subsequent processing. The source of the primary radiation for the two methods is different: X-ray photoelectron spectroscopy makes use of soft X-rays, generally AlKα or MgKα, whereas AES and SAM rely on the use of an electron gun. The specification for electron guns used in Auger analysis varies tremendously, particularly as far as the spatial resolution is concerned which, for finely

---

[1]Units: in electron spectroscopy, energies are expressed in the non-SI unit the electron volt. The conversion factor to the appropriate SI unit is $1\,eV = 1.595 \times 10^{-19}\,J$.

focused guns, may be between 5 μm and <10 nm. In principle, the same energy analyser may be used for both XPS and AES; consequently, the two techniques are often to be found in the same analytical instrument.

Before considering the uses and applications of the two methods, a brief review of the basic physics of the two processes and the strengths and weaknesses of each technique will be given.

1.2 Notation

XPS and AES measure the energy of electrons emitted from a material. It is necessary, therefore, to have some formalism to describe which electrons are involved with each of the observed transitions. The notation used in XPS is different from that used in AES. XPS uses the so-called spectroscopists' or chemists' notation while Auger electrons are identified by the X-ray notation.

1.2.1 Spectroscopists' notation

In this notation the photoelectrons observed are described by means of their quantum numbers. Transitions are usually labelled according to the scheme nl_j. The first part of this notation is the principal quantum number, n. This takes integer values of 1, 2, 3 etc. The second part of the nomenclature, l, is the quantum number which describes the orbital angular momentum of the electron. This takes integer values 0, 1, 2, 3 etc. However, this quantum number is usually given a letter rather than a number as shown in Table 1.1.

Table 1.1 Notation given to the quantum numbers which describe orbital angular momentum

Value of l	Usual notation
0	s
1	p
2	d
3	f

The peaks in XPS spectra, derived from orbitals whose angular momentum quantum number is greater than 0, are usually split into two. This is a result of the interaction of the electron angular momentum due to its spin with its orbital angular momentum. Each electron has a quantum number associated with its spin angular momentum, s^2. The value of s can be either $+1/2$ or $-1/2$. The two angular momenta are added vectorially to produce the quantity j in the expression nl_j, i.e., $j = |l + s|$. Thus, an electron from a p orbital can have a j value of $1/2$ $(l - s)$ or $3/2$ $(l + s)$; similarly, electrons from a d orbital can have j values of either $3/2$ or $5/2$. The relative intensity of the components of the doublets formed by the spin orbit coupling is dependent upon their relative populations (degeneracies) which are given by the expression $(2j + 1)$ so, for an electron from a d orbital, the relative intensities of the $3/2$ and $5/2$ peaks are 2:3. The spacing between the components of the doublets depends upon the strength of the spin orbit coupling. For a given value of both n and l the separation increases with the atomic number of the atom. For a given atom, it decreases both with increasing n and with increasing l.

Figure 1.1 shows an XPS spectrum from Sn with the peaks labelled according to this notation and illustrating the splitting observed in the

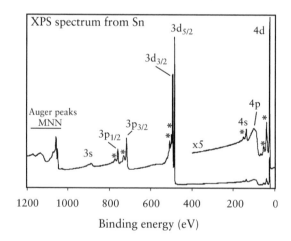

Figure 1.1 Survey spectrum from Sn showing the XPS transitions accessible using AlKα radiation, the features marked with an asterisk are electron energy loss features due to plasmon excitation

---

[2]The electron spin quantum number, s, should not be confused with the description of the orbitals whose angular momentum is equal to zero.

peaks due to electrons in 3p and 3d orbitals while splitting in the 4d and 4p peaks is too small to be observed.

1.2.2 X-ray notation

In X-ray notation, the principal quantum numbers are given letters K, L, M, etc. while subscript numbers refer to the j values described above. The relationship between the notations is given in Table 1.2.

Table 1.2 The relationship between quantum numbers, spectroscopists' notation and X-ray notation

Quantum numbers				Spectroscopists' notation	X-ray notation
n	l	s	j		
1	0	$+1/2, -1/2$	1/2	$1s_{1/2}$	K
2	0	$+1/2, -1/2$	1/2	$2s_{1/2}$	L_1
2	1	$+1/2$	1/2	$2p_{1/2}$	L_2
2	1	$-1/2$	3/2	$2p_{3/2}$	L_3
3	0	$+1/2, -1/2$	1/2	$3s_{1/2}$	M_1
3	1	$+1/2$	1/2	$3p_{1/2}$	M_2
3	1	$-1/2$	3/2	$3p_{3/2}$	M_3
3	2	$+1/2$	3/2	$3d_{3/2}$	M_4
3	2	$-1/2$	5/2	$3d_{5/2}$	M_5
					etc.

As will be seen later, the Auger process involves three electrons and so the notation has to take account of this. This is done simply by listing the three electrons; a peak in an Auger spectrum may be labelled, for example, KL_1L_3 or $L_2M_5M_5$. For convenience, the subscripts are sometimes omitted.

1.3 X-ray Photoelectron Spectroscopy (XPS)

In XPS we are concerned with a special form of photoemission, i.e., the ejection of an electron from a core level by an X-ray photon of energy $h\nu$. The energy of the emitted photoelectrons is then analysed by the electron spectrometer and the data presented as a graph of intensity (usually expressed as counts or counts/s) versus electron energy – the X-ray induced photoelectron spectrum.

The kinetic energy (E_K) of the electron is the experimental quantity measured by the spectrometer, but this is dependent on the photon energy of the X-rays employed and is therefore not an intrinsic material property. The binding energy of the electron (E_B) is the parameter which identifies the electron specifically, both in terms of its parent element and atomic energy level. The relationship between the parameters involved in the XPS experiment is:

$$E_B = h\nu - E_K - W$$

where $h\nu$ is the photon energy, E_K is the kinetic energy of the electron, and W is the spectrometer work function.

As all three quantities on the right-hand side of the equation are known or measurable, it is a simple matter to calculate the binding energy of the electron. In practice, this task will be performed by the control electronics or data system associated with the spectrometer and the operator merely selects a binding or kinetic energy scale whichever is considered the more appropriate.

The process of photoemission is shown schematically in Figure 1.2, where an electron from the K shell is ejected from the atom (a 1s photoelectron). The photoelectron spectrum will reproduce the electronic structure of an element quite accurately since all electrons with a binding

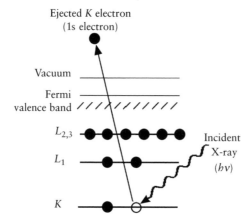

Figure 1.2 Schematic diagram of the XPS process, showing photoionization of an atom by the ejection of a 1s electron

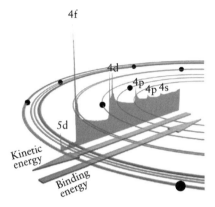

Figure 1.3 Photo electron spectrum of lead showing the manner in which electrons escaping from the solid can contribute to discrete peaks or suffer energy loss and contribute to the background; the spectrum is superimposed on a schematic of the electronic structure of lead to illustrate how each orbital gives rise to photoelectron lines

energy less than the photon energy will feature in the spectrum. This is illustrated in Figure 1.3 where the XPS spectrum of lead is superimposed on a representation of the electron orbitals. Those electrons which are excited and escape without energy loss contribute to the characteristic peaks in the spectrum; those which undergo inelastic scattering and suffer energy loss contribute to the *background* of the spectrum. Once a photoelectron has been emitted, the ionized atom must relax in some way. This can be achieved by the emission of an X-ray photon, known as X-ray fluorescence. The other possibility is the ejection of an Auger electron. Thus Auger electrons are produced as a consequence of the XPS process often referred to as X-AES (X-ray induced Auger electron spectroscopy). X-AES, although not widely practised, can yield valuable chemical information about an atom. For the time being we will restrict our thoughts to AES in its more common form, which is when a finely focused electron beam causes the emission of Auger electrons.

1.4 Auger Electron Spectroscopy (AES)

When a specimen is irradiated with electrons, core electrons are ejected in the same way that an X-ray beam will cause core electrons to be

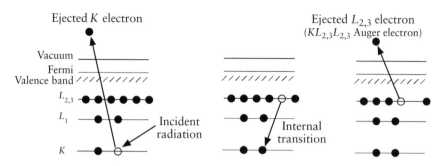

Figure 1.4 Relaxation of the ionized atom of Figure 1.2 by the emission of a $KL_{2,3}L_{2,3}$ Auger electron

ejected in XPS. The difference is that in the case of electron irradiation the secondary electrons contain no analytical information – although those of low energy are very useful for imaging purposes as in scanning electron microscopy. However, once an atom has been ionized it must, in some way, return to its ground state. The emission of an X-ray photon may occur, which is the basis of electron probe microanalysis (EPMA), carried out in many electron microscopes by either energy dispersive (EDX) or wavelength dispersive (WDX) spectrometers. The other possibility is that the core hole (for instance a K shell vacancy as shown in Figure 1.2) may be filled by an electron from a higher level, the $L_{2,3}$ level in Figure 1.4. In order to conform with the principle of the conservation of energy, another electron must be ejected from the atom, e.g., another $L_{2,3}$ electron in the schematic of Figure 1.4. This electron is termed the $KL_{2,3}L_{2,3}$ Auger electron.

It is common to omit the subscripts when referring to the group of Auger emissions involving the same principal quantum numbers, for example the term Si KLL is used to refer to the whole group of KLL emissions from silicon. Similar generalizations can be used for emissions involving core or valence electrons. It is not uncommon to see terms such as NVV, which refer to Auger emissions in which an electron is removed from the N orbital to be replaced by an electron from the valence shell causing a second valence electron to be emitted. Even more general is the term CVV in which the C refers to any core electron, using this approach a CCC transition indicates the involvement of three electrons from core levels. In general, it is the CCC Auger transitions which provide chemical information in Auger electron spectroscopy.

The kinetic energy of a $KL_{2,3}L_{2,3}$ Auger electron is approximately equal to the difference between the energy of the core hole and the energy levels of the two outer electrons, $E_{L_{2,3}}$ (the term $L_{2,3}$ is used in this case because, for light elements, L_2 and L_3 cannot be resolved):

$$E_{KL_{2,3}L_{2,3}} \approx E_K - E_{L_{2,3}} - E_{L_{2,3}}$$

This equation does not take into account the interaction energies between the core holes ($L_{2,3}$ and $L_{2,3}$) in the final atomic state nor the inter- and extra-relaxation energies which come about as a result of the additional core screening needed. Clearly, the calculation of the energy of Auger electron transitions is much more complex than the simple model outlined above, but there is a satisfactory empirical approach which considers the energies of the atomic levels involved and those of the next element in the periodic table.

Following this empirical approach, the Auger electron energy of transition $KL_1L_{2,3}$ for an atom of atomic number Z is written:

$$E_{KL_1L_{2,3}}(Z) = E_K(Z) - 1/2[E_{L_1}(Z) + E_{L_1}(Z+1)] \\ - 1/2[E_{L_{2,3}}(Z) + E_{L_{2,3}}(Z+1)]$$

Clearly for the $KL_{2,3}L_{2,3}$ transition the second and third terms of the above equation are identical and the expression is simplified to:

$$E_{KL_{2,3}L_{2,3}}(Z) = E_K(Z) - [E_{L_{2,3}}(Z) + E_{L_{2,3}}(Z+1)]$$

It is the kinetic energy of this Auger electron ($E_{KL_{2,3}L_{2,3}}$) that is the characteristic material quantity irrespective of the primary beam composition (i.e., electrons, X-rays, ions) or its energy. For this reason Auger spectra are always plotted on a kinetic energy scale.

The use of a finely focused electron beam for AES enables us to achieve surface analysis at a high spatial resolution, in a manner analogous to EPMA in the scanning electron microscope. By combining an electron spectrometer with an ultra-high vacuum (UHV) SEM it becomes possible to carry out scanning Auger microscopy. In this mode of operation various imaging and chemical mapping procedures become possible.

1.5 Scanning Auger Microscopy (SAM)

In the scanning Auger microscope various modes of operation are available, the variable quantities being the position of the electron probe on the specimen (x and y) and the setting of the electron energy analyser (E) corresponding to the energy of emitted electrons to be analysed. The various possibilities are summarized in Table 1.3.

As the Auger electron yield is very sensitive to the electron take-off angle, an image of Auger electron intensities will invariably reflect the surface topography of the specimen, possibly more strongly than the chemical variations, as illustrated (Figure 1.5) in the Auger map of carbon fibres (Figure 1.5(b)) which is very similar to the SEM image

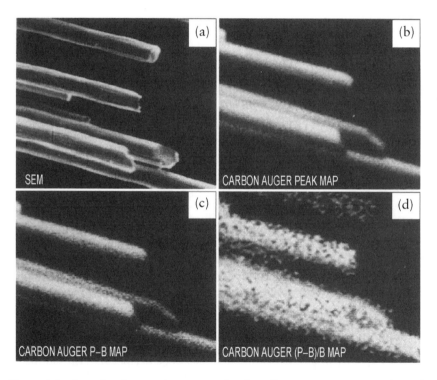

Figure 1.5 Scanning Auger microscopy of carbon fibres: (a) SEM image, (b) peak map (P) of carbon Auger electrons, (c) peak-background map (P − B), B recorded 40 eV from Auger peak, (d) correction for topographic effects using (P − B)/B algorithm; the diameter of the fibres is 7 μm

Table 1.3 Modes of analysis available with SAM

Mode of analysis	Scanned	Fixed
Point analysis	E	x, y
Line scan	x	E, y
Chemical map	x, y	E

(Figure 1.5(a)). The problem is overcome by recording a background (B) as well as the Auger peak (P) map. However, a simple subtraction of the background counts from the peak intensity (P − B) is not sufficient as shown by the (P − B) map of Figure 1.5(c). The use of a simple algorithm such as (P − B)/B, allows correction for the effects of surface topography (Figure 1.5(d)) where variation in intensity due to the cylindrical shape of the fibres has been completely suppressed and only chemical information remains.

1.6 The Depth of Analysis in Electron Spectroscopy

The depth of analysis in both XPS and AES varies with the kinetic energy of the electrons under consideration. It is determined by a quantity known as the attenuation length (λ) of the electrons, which is related to the inelastic mean free path (IMFP). It varies as $E^{0.5}$ in the energy range of interest in electron spectroscopy and various relationships have been suggested which relate λ to electron energy and material properties. One such equation proposed by Seah and Dench (1979) of the National Physical Laboratory, UK, is given below:

$$\lambda = \frac{538a_A}{E_A^2} + 0.41a_A(a_AE_A)^{0.5}$$

where E_A is the energy of the electron in eV, a_A^3 is the volume of the atom in nm^3 and λ is in nm.

Values for IMFP may be derived using optical spectroscopy, or reflection electron energy loss spectroscopy (REELS), those for λ are generally deduced from XPS and Auger spectroscopy measurements. In general, the attenuation length is about 10 per cent less than the IMFP. Various databases exist from which values of IMFP and attenuation length can be obtained.

The intensity of electrons (I) emitted from all depths greater than d in a direction normal to the surface is given by the Beer–Lambert relationship:

$$I = I_0 \exp(-d/\lambda)$$

where I_0 is the intensity from an infinitely thick, uniform substrate. For electrons emitted at an angle θ to the surface normal, this expression becomes:

$$I = I_0 \exp(-d/\lambda\cos\theta)$$

The variation of electron intensity with depth is shown schematically, for a carbon substrate, in Figure 1.6.

The Beer–Lambert equation can be manipulated in a variety of ways to provide information about overlayer thickness and to provide a non-destructive depth profile (i.e., without removing material by mechanical, chemical or ion-milling methods). Using the appropriate analysis of the above equation, it can be shown that by considering electrons that emerge at 90° to the sample surface, some 65 per cent of the signal in

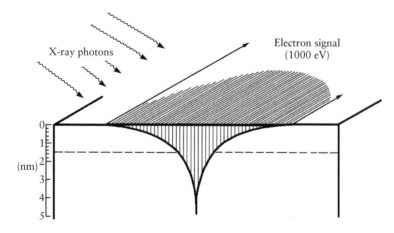

Figure 1.6 Electron intensity as a function of depth, the horizontal dashed line indicates a distance from the surface of the attenuation length (λ)

electron spectroscopy will emanate from a depth of less than λ, 85 per cent from a depth of $<2\lambda$, and 95 per cent from a depth of $<3\lambda$, as illustrated in Figure 1.6. The values of the inelastic mean free paths are of the order of a few nanometers and this is the mechanism by which the signal from the vast majority of the atoms present in the sample is filtered out (stated as a requirement for a surface analysis technique in Section 1.1).

However, the depth from which information can be derived is a few nanometers, a fact which can be exploited in angle resolved measurements to obtain compositional depth profiles. This will be discussed in more detail later.

1.7 Comparison of XPS and AES/SAM

Although it is difficult to make a comparison of techniques before they are described and discussed in detail, it is pertinent at this point to outline the strengths and weaknesses of each to provide background information.

XPS is also known by the acronym ESCA (electron spectroscopy for chemical analysis). It is this *chemical* specificity which is the major strength of XPS as an analytical technique, one for which it has become deservedly popular. By this we mean the ability to define not only the elements present in the analysis but also the chemical state. In the case of iron, for instance, the spectra of Fe^0, Fe^{2+}, and Fe^{3+} are all slightly different and to the expert eye are easily distinguishable. However, such information is attainable in XPS only at the expense of spatial resolution, and XPS is usually regarded as an area averaging technique. Small area XPS (SAXPS or even SAX) is available on most modern instruments and, when operating in this mode, a spectroscopic spatial resolution of about 10 μm is possible. Many modern instruments offer imaging XPS and such imaging may have a spatial resolution of <3 μm. Set beside a spatial resolution of 10 to 15 nm which can be achieved on the latest commercial Auger microprobes, it becomes clear that the XPS is not the way to proceed for surface analysis at very high spatial resolution, but the advantages of the levels of information available from an XPS analysis (the ease with which a quantitative analysis can be achieved, its applicability to insulators, and the ready availability of chemical state information) will often offset this.

In addition to the chemical state information referred to above, XPS spectra can be quantified in a very straightforward manner and meaningful comparisons can be made between specimens of a similar type. Quantification of Auger data is rather more complex and the accuracy obtained is generally not so good. Because of the complementary nature of the two methods and the ease with which Auger and photoelectron analyses can be made on the same instrument, the two methods have come to be regarded as the most important methods of surface analysis in the context of materials science. All manufacturers of electron spectrometers offer both XPS and AES options for their systems.

1.8 The Availability of Surface Analytical Equipment

The capital cost of an XPS/AES/SAM spectrometer is high when compared with most electron microscopes, and at the time of writing is of the order of £0.3 million to £0.5 million for a comprehensive system. This, allied to the fairly steep learning curve that the newcomer must ascend before confidence in the technique is obtained, has lead to the development of laboratories offering surface analysis as a service facility; such laboratories may be found throughout the world and they are often associated with universities but the balance between academic and industrial work varies greatly. The use of service facilities presents a very attractive proposition to inexperienced users in that expert advice is always on hand to ensure the efficient use of instrument time – a factor that is of paramount importance since the daily charge for the use of such a facility can exceed £1500. It is not unusual for analysts to 'cut their teeth' on the field of surface analysis in such a way; once the need within their own company (and their own personal expertise) has been established, a surface analysis system can then be specified for their own particular needs.

Although originally the exclusive preserve of research laboratories and academic institutions, surface analysis facilities are now frequently to be found in trouble-shooting and quality assurance roles. As the techniques find wider applications, so the market grows and manufacturers are very willing to continue developing their spectrometers. Thus, the future for XPS and AES seems assured well into the future.

2 Electron Spectrometer Design

The design and construction of electron spectrometers is a very complex undertaking and will usually be left to one of the handful of specialist manufacturers worldwide, although many users specify minor modifications to suit their own requirements. The various modules necessary for analysis by electron spectroscopy are (in addition to a specimen): a source of the primary beam (either X-rays or electrons); an electron energy analyser and detection system, all contained within a vacuum chamber; a data system which is nowadays an integral part of the system.

2.1 The Vacuum System

All commercial spectrometers are now based on vacuum systems designed to operate in the ultra-high vacuum (UHV) range of 10^{-8} to 10^{-10} mbar[3], and it is now generally accepted that XPS and AES experiments must be carried out in this pressure range. The reasons for this are as follows.

1. The analytical signal of low-energy electrons is easily scattered by the residual gas molecules and, unless their concentration is kept to an acceptable level, the total spectral intensity will decrease, while the noise present within the spectrum will increase.

---

[3]Units: in electron spectroscopy, pressures are expressed in the non-SI unit the mbar. The conversion factor to the appropriate SI unit is 1 mbar $= 10^2$ Pa.

2. More importantly, the UHV environment is necessary because of the surface sensitivity of the techniques themselves. At 10^{-6} mbar, it is possible for a monolayer of gas to be adsorbed onto a solid surface in about 1 s. This time period is short compared with that required for a typical spectral acquisition, clearly establishing the need for a UHV environment during analysis.

The manner in which such a vacuum is established will depend on customer and manufacturer preferences. The chambers and associated piping will invariably be made of stainless steel and joints will usually be effected by using flanges, equipped with knife-edges, which are tightened onto copper gaskets (a system generally referred to as *conflat*, following the designation by Varian Associates who own the trademark, but which are now supplied by a number of manufacturers).

UHV conditions are usually obtained in a modern electron spectrometer using ion pumps. Turbomolecular pumps are popular with some users, especially if it is necessary to pump large quantities of noble gases. Diffusion pumps, which were very popular some time ago, have now largely disappeared from modern commercial instruments. Whichever type of pump is chosen, it is common to use a titanium sublimation pump to assist the primary pumping and to achieve the desired vacuum level. All UHV systems need baking from time to time to remove adsorbed layers from the chamber walls, the baking temperature is dictated by the analytical options fitted to the spectrometer but is usually in the range 100–160°C for routine use.

The trajectory of the electrons is strongly influenced by the Earth's magnetic field. Consequently, some form of magnetic screening is required around the sample and electron analyser. There are two approaches to this problem. The most elegant solution is to fabricate the entire analysis chamber from a material with high magnetic permeability (μ-metal). An acceptable alternative is to fabricate shielding panels, either as sleeving within the instrument or as a bolt-on outer shroud. The methodology depends on the manufacturer. In addition, compensation coils may be arranged around the analyser and transfer lens to mitigate the effect of such magnetic fields.

2.2 The Sample

Although electron spectroscopy can be carried out successfully on gases and liquids as well as solids, gases and liquids necessarily yield bulk molecular and chemical information rather than surface chemical information. Consequently, we shall restrict our discussions here to solid samples. The criteria for analysis by AES and XPS are not the same, the requirements for specimens for Auger spectroscopy being somewhat more stringent.

Samples for both XPS and AES must be stable within the UHV chamber of the spectrometer. Very porous materials (such as some ceramic and polymeric materials) can pose problems as well as those which either have a high vapour pressure or have a component which has a high vapour pressure (such as a solvent residue). In this context, 10^{-7} mbar is considered a high vapour pressure.

As far as XPS is concerned, once these requirements have been fulfilled the sample is amenable to analysis. For Auger analysis, however, the use of an electron beam dictates that for routine analysis the specimen should be conducting and effectively earthed in addition to the vacuum compatibility requirements outlined above. As a guide, if a specimen can be imaged (in an uncoated condition) in an SEM without any charging problems, a specimen of similar type can be analysed by Auger electron spectroscopy. The analysis of insulators such as polymers and ceramics by AES is quite feasible but its success relies heavily on the skill and experience of the instrument operator. Such analysis is achieved by ensuring that the incoming beam current is exactly balanced by the combined current of emitted electrons (all secondaries including Auger electrons, backscattered and elastically scattered electrons, etc.), by adjusting the beam energy (3–5 keV), specimen current (very low probably <10 nA) and electron take-off angle. With the recent introduction of ion guns capable of beam energies below ~50 eV, it is now possible to obtain high-quality Auger data from insulators by the simple expedient of using a low flux of such ions for charge control. The positive ions neutralize the surface and their energy is too low to cause atoms to be sputtered from the surface. In the special case of a thin insulating layer on a conducting or semiconducting sample, the use of a high primary beam energy can induce a conducting track within the insulating layer and excess charge can be dissipated.

The mounting of conducting samples is best achieved with clips or bolt-down assemblies. For XPS the use of double-sided adhesive tape can be used but only sparingly because it can have mobile materials, such as release agents, on its surface which can contaminate the surface under investigation.

For conducting specimens, a fine stripe of conducting paint, in addition to the adhesive tape, is all that is necessary to prevent sample charging. Solvents in the conductive paint can cause the pump down time to be extended as they evaporate into the vacuum. Alternatively, metal tape with a metal loaded (conducting) adhesive may be used. Most laboratories have a selection of sample holders, usually fabricated in-house, to accommodate large and awkwardly shaped specimens. Discontinuous specimens present rather special problems. In the case of powders, the best method is embedding them in indium foil, but if this is not feasible, dusting them onto double-sided adhesive tape can be a very satisfactory alternative. Fibres and ribbons can be mounted across a gap in a specimen holder ensuring that no signal from the mount is detected in the analysis.

The type of sample mount varies with the instrument design and most modern spectrometers use a sample stub similar to the type employed in scanning electron microscopy, or a sample platter that will accommodate many samples. For analysis, the sample is held in a high-resolution manipulator with x, y, and z translations, and tilt and rotation about the z-axis (azimuthal rotation). For scanning Auger microscopy, where the time taken to acquire high-resolution maps can be about 1 h, the stability of the stage is critical, since any drift during analysis will degrade the resolution of the images. Image registration software, used during acquisition, can mitigate the effects of a small amount of drift. For angle resolved XPS (ARXPS), the amount of backlash in the rotary drive must be small and the scale should be graduated in increments of 1° for manual operation.

Once mounted for analysis, heating or cooling of the specimen can be carried out *in vacuo*. Cooling is generally restricted to liquid nitrogen temperatures although liquid helium stages are available. Heating may be achieved by direct (contact) heating using a small resistance heater or by electron bombardment for higher temperatures. Such heating and cooling will either be a preliminary to analysis or carried out during the analysis itself (with the obvious exception of electron bombardment heating). Heating in particular will often be carried out in a preparation chamber because of the possibility of severe outgassing encountered at higher temperatures.

The routine analysis of multiple similar specimens by AES and, in particular, XPS can be a time-consuming business and some form of automation is desirable. This is available from several manufacturers in the form of a computer-driven carousel or table which enables a batch of specimens to be analysed when a machine is left unattended, typically overnight. A modern commercial electron spectrometer is illustrated in Figure 2.1.

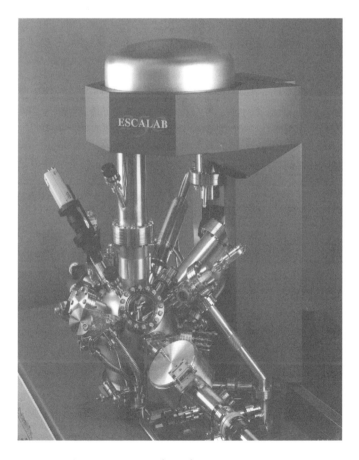

Figure 2.1 A modern electron spectrometer

2.3 X-ray Sources for XPS

2.3.1 The twin anode X-ray source

X-rays are generated by bombarding an anode material with high-energy electrons. The electrons are emitted from a thermal source, usually in the form of an electrically heated tungsten filament but, in some focusing X-ray monochromators, a lanthanum hexaboride emitter is used because of its higher current density (brightness). The efficiency of X-ray emission from the anode is determined by the electron energy, relative to the X-ray photon energy. For example, the AlKα (energy 1486.6 eV) photon flux from an aluminium anode increases by a factor of more than five if the electron energy is increased from 4 keV to 10 keV. At a given energy, the photon flux from an X-ray anode is proportional to the electron current striking the anode. The maximum anode current is determined by the efficiency with which the heat, generated at the anode, can be dissipated. For this reason, X-ray anodes are usually water-cooled.

The choice of anode material for XPS determines the energy of the X-ray transition generated. It must be of high enough photon energy to excite an intense photoelectron peak from all elements of the periodic table (with the exception of the very lightest); it must also possess a natural X-ray line width that will not broaden the resultant spectrum excessively. The most popular anode materials are aluminium and magnesium. These are usually supplied in a single X-ray gun with a twin anode configuration providing AlKα or MgKα photons of energy 1486.6 eV and 1253.6 eV respectively. This is possible because, unlike X-ray diffraction (XRD) anodes, it is the anode and not the filament which is at a high potential (for XRD the filament is at a high negative potential and the anode at ground; for XPS the filament is at or near ground and the anode at a high positive potential of 10–15 kV).

Such twin anode assemblies are useful as they provide a modest depth profiling capability – the difference in the analysis depth of organic materials of the carbon 1s electron is about 1 nm greater for electrons excited by AlKα, (analysis depth approximately $7\cos\theta$ nm) compared with MgKα (analysis depth approximately $6\cos\theta$ nm). More importantly, they provide the ability to differentiate between Auger and photoelectron transitions when the two overlap in one radiation. XPS peaks will change to a position 233 eV higher on a kinetic energy scale on switching from MgKα to AlKα whereas the energy of Auger transitions remains constant. On a binding energy scale, of course, the reverse is true as shown in Figure 2.2.

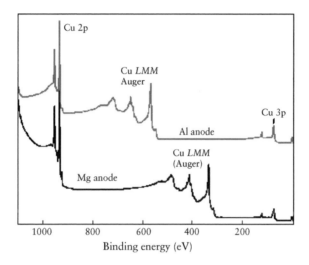

Figure 2.2 Comparison of XPS spectra recorded from copper using AlKα (upper) and MgKα (lower) radiation; note that on a binding energy scale the XPS peaks remain at constant values but the X-AES transitions move by 233 eV on switching between the two sources

Table 2.1 Possible anode materials for XPS

Element	Line	Energy (eV)	Full-width half maximum (eV)
Y	$M\zeta$	132.3	0.47
Zr	$M\zeta$	151.4	0.77
Mg	$K\alpha_{1,2}$	1253.6	0.7
Al	$K\alpha_{1,2}$	1486.6	0.9
Si	$K\alpha$	1739.6	1.0
Zr	$L\alpha$	2042.4	1.7
Ag	$L\alpha$	2984.4	2.6
Ti	$K\alpha$	4510.9	2.0
Cr	$K\alpha$	5417.0	2.1

The photon energies and peak widths of MgKα and AlKα are compared with those of other elements in Table 2.1. In a twin anode arrangement, any two of these anode materials can be used in any combination. There are two advantages of higher-energy anodes.

1. Energy levels not available in conventional XPS become accessible – in AlKα radiation the Mg 1s electron is the highest K electron attainable, in SiKα this is extended to the Al 1s electron, in ZrLα the Si 1s electron, in AgLα the Cl 1s electron and in TiKα the Ca 1s electron.

2. Because the use of higher-energy photon sources increases the kinetic energy of the ejected photoelectrons available when compared with conventional XPS, higher-energy XPS provides a non-destructive means of increasing the analysis depth. It is therefore possible to build up a depth profile of a specimen merely by changing the X-ray source and monitoring the apparent change in composition.

2.3.2 X-ray monochromators

Increased performance of monochromators, accompanied by improved sensitivity of modern spectrometers, means that analysis using monochromatic X-rays is becoming much more common. Indeed, some commercial spectrometers are equipped with a monochromator as their only X-ray source.

The purpose of an X-ray monochromator is to produce a narrow X-ray line by using diffraction in a crystal lattice; Figure 2.3 illustrates the

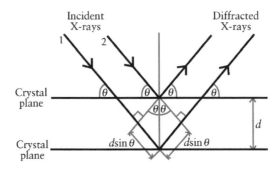

Figure 2.3 Diffraction of X-rays at a quartz crystal

Table 2.2 Radiation which can be produced using a quartz crystal monochromator; note that in the case of the CrKα line the wavelength does not fulfil the Bragg condition and the weaker, satellite line CrKβ must be used

Diffraction order	X-ray line	Energy (eV)
1	AlKα	1486.6
2	AgLα	2984.3
3	TiKα	4510.0
4	CrKβ	5946.7

process. X-rays strike the parallel crystal planes at an angle θ and are reflected at the same angle. The distance travelled by the X-rays depends upon the crystal plane at which they are reflected. Figure 2.3 shows two adjacent crystal planes with X-rays being reflected from each. If the distance between the planes is d then the difference in the path length is $2d\sin\theta$. If this distance is equal to an integral number of wavelengths then the X-rays interfere constructively, if not destructive interference takes place. This gives rise to the well-known Bragg equation:

$$n\lambda = 2d\sin\theta$$

in which $n=$ the diffraction order, $\lambda=$ the X-ray wavelength, $d=$ the crystal lattice spacing, and $\theta=$ Bragg angle (angle of diffraction).

At present, all commercially available X-ray monochromators used for XPS employ a quartz crystal (usually the $(10\bar{1}0)$ crystal face) as the diffraction lattice. Usually, the monochromator on an XPS instrument is used for AlKα radiation, the spacing in the quartz crystal lattice means that first-order reflection occurs at a convenient angle. However, other materials and other diffraction orders have been used, as shown in Table 2.2.

Quartz is a convenient material because it is relatively inert, compatible with UHV conditions, it can be bent and/or ground into the correct shape and its lattice spacing provides a convenient diffraction angle for AlKα radiation.

There exist a number of reasons for choosing to use an X-ray monochromator on an XPS spectrometer.

1. The primary reason for using monochromated radiation is the reduction in X-ray line width, for example, from 0.9 eV to

approximately 0.25 eV for AlKα, and from 2.6 eV to 1.2 eV for AgLα. Narrower X-ray line width results in narrower XPS peaks and consequently better chemical state information.

2. Unwanted portions of the X-ray spectrum, i.e., satellite peaks and the bremsstrahlung continuum are also removed.

3. For maximum sensitivity, a twin anode X-ray source is usually positioned as close to the sample as possible. The sample is therefore exposed to the radiant heat from the source region which could damage or alter the surface of delicate samples. When a mono-chromator is used, this heat source is remote from the sample and thermally induced damage is avoided.

4. It is possible to focus X-rays into a small spot using the mono-chromator. This means that small area XPS can be conducted with high sensitivity.

5. Use of a focusing monochromator means that only the area of the specimen being analysed is exposed to X-rays. Thus, a number of samples may be loaded into the spectrometer without the risk of X-rays damaging samples while they await analysis. Similarly, multi-point analysis can be performed on the same delicate sample.

Some of these advantages are illustrated in Figure 2.4. This shows the XPS spectrum of Ag 3d electrons acquired using monochromatic and non-monochromatic X-rays. The analyser was set to the same con-ditions for each spectrum. There is a clear difference in the peak width, the background is higher using the non-monochromatic X-rays and X-ray satellites are clearly visible when the non-monochromatic source is used.

By using a focusing X-ray monochromator, illustrated in Figure 2.5, it is possible to produce a small area XPS (SAX) analysis and this now forms the basis of commercial instruments with a spatial resolu-tion of <15 μm. This is one route to SAX; the other commercially available method, electron-optical aperturing, is discussed later in this chapter.

In Figure 2.5, the quartz crystal is curved in such a way that it focuses the X-ray beam as well as causing it to be diffracted. By this means, the

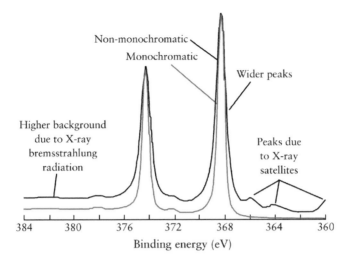

Figure 2.4 A comparison of the Ag 3d spectra acquired with monochromatic and non-monochromatic X-rays (the spectra are normalized to the maximum peak intensity)

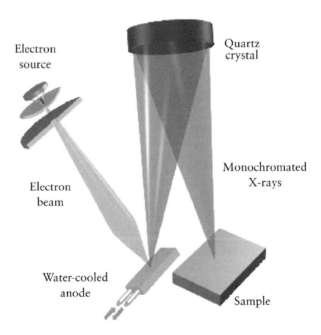

Figure 2.5 Schematic diagram of a focusing X-ray monochromator

size of the X-ray spot on the surface of the sample is approximately equal to that of the electron spot on the anode. Thus, by varying the focusing of the electron source, the analyst can vary the analysis area.

2.3.3 Charge compensation

Photoemission from an insulating sample causes electrostatic charging to occur in the positive direction. This results in a shift in the peak position in the direction of higher binding energy. When the photoemission is excited by a non-monochromatic X-ray source, there are usually a sufficient number of low-energy electrons available in the neighbourhood of the sample to effectively neutralize the sample and allow high-quality XPS spectra to be obtained.

When monochromatic X-ray sources are used, these low-energy electrons are not produced in such large numbers near the sample and so neutralization does not take place. Indeed, because the X-ray line width from a monochromator is much narrower, the need for effective charge compensation is even greater. The charge compensation must also be uniform over the analysis area to prevent broadening of the XPS peaks.

When charge compensation is necessary, it is normal practice to flood the sample with low-energy electrons. It is not usual to attempt to balance the charging exactly, an excess of electrons is used to produce a uniform negative charge of known magnitude to be produced at the surface of the sample. The peaks can then be shifted to their correct positions during data processing. This technique minimizes the risk of differential or non-uniform charging.

The electron beam used for charge compensation should be of low energy to avoid the risk of damage to the surface of the sample but must be of a high enough flux to adequately compensate for the charging. Typically, the energy of the electrons is less than 5 eV.

2.4 The Electron Gun for AES

Since 1969 (the year in which Auger electron spectrometers became commercially available), AES has become a well-accepted analytical method for the provision of surface analyses at high spatial resolution. The intervening years have seen the electron guns improve from the 500 μm resolution of the converted oscilloscope gun of the late 1960s to the <10 nm resolution which is typical of top-of-the range Auger microprobes of today. In between these two extremes are the 5 μm and 100 nm guns which are typical of electron guns in use with instruments on which Auger is not the prime technique (multi-technique instruments).

The critical components of the electron gun are the electron source and the lens assemblies for beam focusing, shaping and scanning. Electron sources may be either thermionic emitters or field emitters, lenses for the electron gun may be electrostatic or, for high-resolution applications, electromagnetic.

If we consider the lenses first, the criterion, which distinguishes an electron gun for AES from one conventional in electron microscopy, is the need to operate in a UHV environment. In the early days of AES, this effectively precluded the use of electromagnetic lenses, as the coils were not able to withstand UHV bakeout temperatures. Consequently, electrostatic lenses were much favoured and by gradual design improvement the stage has been reached where electron guns with electrostatic lenses can achieve a spatial resolution of <100 nm. The major step forward, as far as scanning Auger microscopy was concerned, was the development of a gun with bakeable electromagnetic lenses; by the late 1970s such lenses, with a spatial resolution of 50 nm and a thermionic emitter, became routinely available. Nowadays, the combination of electromagnetic lenses and a field emission source of electrons provides Auger instruments whose electron spot size can be less than 10 nm.

2.4.1 Electron sources

To be useful as an electron source for Auger electron spectroscopy a source should have the following properties.

1. *Stability*. The current emitted from the source should be highly stable over long periods. Although spectra can be obtained in a few

minutes, if the spectrometer is to be used for depth profiling then stability over many hours is essential.

2. *Brightness*. High emission currents from a small emitted area are required if the eventual spot size at the specimen is to be small.

3. *Mono-energetic*. The focal length of electromagnetic and electro-static lenses is dependent upon the energy of the electrons. This means that the optimum focusing conditions can only occur for electrons having a very small range of kinetic energy. A wide energy spread will therefore result in a large spot size at the sample.

4. *Longevity*. Under normal operating conditions the emitter must not need to be replaced for many hundreds of hours. Replacement of emitters requires that the vacuum be broken and the instrument be baked before it can be used. This operation means that the instrument cannot be used for a period of 2 or 3 days.

A number of different types of source have been used in commercial instruments; the more important of these will now be described.

Thermionic emitter

The simplest form of thermionic source is a tungsten wire fabricated in the form of a hairpin. When an electric current is passed through it, the temperature rises giving electrons sufficient energy to overcome the work function and be released into free space. The work function is the energy required for an electron to escape from a solid surface; for tungsten this is about 4.5 eV. By reducing the work function, the num-ber of electrons emitted per unit area per unit solid angle can be in-creased, thereby increasing the so-called *brightness* of the source. Shaping the emitter into the form of a hairpin reduces the area of the filament which contributes to the electron beam and therefore minimizes the spot size. This reduction is partly due to geometrical reasons and partly due to the fact that the electrical voltage drop across the emitter area is minimized thus reducing the energy spread.

Simple thermionic emitters are inexpensive and robust but they lack brightness and so it is difficult, using this type of source, to attain spot sizes for Auger analysis below about 200 nm.

Lanthanum hexaboride emitter

A widely used material for high-brightness sources is single crystal lanthanum hexaboride (LaB_6). The source consists of a small, indirectly heated, crystal of LaB_6. The crystal is cylindrical in shape and conical at one end. The tip of the cone is ground to form a flat surface of about 15 μm in diameter and it is important that the flat surface exposes the $\langle 100 \rangle$ crystal face.

This material has a much lower work function than tungsten (2.6 eV compared with 4.5 eV), which means that it has a high emission density even at a much lower temperature. The operating temperature for an LaB_6 emitter is about 1800 K whereas 2300 K is typical for a tungsten emitter. The lower operating temperature and the fact that there is no voltage drop across the emitting surface mean that the energy spread in the electron beam is small compared with that from a tungsten emitter. For stable operation, this type of emitter requires a better vacuum than the tungsten emitter.

Cold field emitter

A type of emitter which has found use in Auger microscopy is that based on the field emission source. The operating principle of a field emission source is not to give the electrons sufficient energy to jump the work function barrier (as in the thermionic process) but to reduce the magnitude of the barrier itself, both in height (marginally) but, more importantly, in width. It is the latter factor which leads to improved electron emission, as electrons from near the Fermi level can penetrate the barrier by quantum mechanical tunnelling and thus escape from the emitter with no loss in energy.

Another characteristic of the field emission source is its narrow energy distribution, as there are no electrons above the Fermi level and those below it have a rapidly decreasing probability of escape. In practical terms, this feature means a smaller spot size because chromatic aberrations within the electron lenses are small.

Usually, field emission is achieved by the application of a very large electrostatic field between the emitter, which itself must be in the form of a needle with a tip radius of approximately 50 nm, and an extraction electrode. The small area of emission from the tip into a small solid angle provides a high brightness compared with thermionic sources, although the total current will be somewhat lower. The emitter material usually employed is a tungsten single crystal.

A cold field emitter in a practical Auger system will provide very small spot sizes and therefore excellent spatial resolution, but suffers from a lack of long-term stability unless operated in an extremely good vacuum ($<10^{-10}$ mbar). The presence of residual gases close to the emitter will have two destabilizing effects.

1. The gases will adsorb on the cold surface contaminating it and affecting its emission current. The emitting area of this type of source is so small that even a single adsorbed molecule on the emitting surface will have a noticeable effect on the emission current. Adsorbed gases can be removed by 'flashing' the tip. This process involves heating the tip to a temperature near its melting point, causing the adsorbed gases to be removed and, if flashing is undertaken in an electrostatic field, a reshaping of the tip. The reshaping is usually beneficial but occasionally it is such that the tip becomes inoperable and must be replaced. Following flashing, emission from the tip will be unstable for a period of about 1 h.

2. Some of the high-energy electrons in the beam will collide with the residual gas molecules forming positive ions. These ions are then accelerated in the electrostatic field and collide with the tip causing sputtering (removal of tungsten atoms from the surface). This affects the radius of the tip and reduces the efficiency of the emission process.

Hot field emitter

This type of electron emitter (also known as a Schottky field emitter) has become very popular in recent years because its brightness is high and stability good. It is, in effect, a combination of a thermionic emitter and a field emitter. The source consists of a single crystal tungsten wire coated with the semiconductor material zirconium oxide. The function of the ZrO_2 is to:

- lower the work function of the emitter and increase the emission current,
- provide a self-cleaning surface, and
- provide a self-healing surface.

Again, the emitting surface is small (about 20 nm in diameter) and so little demagnification is required to achieve the small spot sizes required in high-resolution scanning Auger microscopy. The tip is heated to about 1800 K in a large electrostatic field. The combination of the high temperature and the field causes electrons to be emitted.

The vacuum requirements for this type of gun are much less stringent than is the case for the cold field emitter, although it is usual to provide additional pumping at the source when this type of emitter is used.

Comparison of electron emitters for Auger electron spectroscopy

Table 2.3 compares the important characteristics of the types of emitter considered here.

Table 2.3   Comparison of electron emitters

	Thermionic	LaB_6	Cold field emitter	Schottky emitter
Work function (eV)	4.5	2.7	4.5	2.95
Brightness ($Acm^{-2}srad^{-1}$)	$<10^5$	$\sim10^6$	10^7 to 10^9	$>10^8$
Current into a 10 nm spot	1 pA	10 pA	10 nA	5 nA
Maximum beam current	1 μA	1 μA	20 nA	200 nA
Minimum energy spread	1.5 eV	0.8 eV	0.3 eV	0.6 eV
Operating temperature (K)	2700	2000	300	1800
Short-term stability	$<1\%$	$<1\%$	$>5\%$	$<1\%$
Long-term stability	High	High	$>10\%/h$	$<1\%/h$
Vacuum required (mbar)	$<10^{-4}$	$<10^{-6}$	$<10^{-10}$	$<10^{-8}$
Typical lifetime (h)	<200	~1000	>2000	>2000
Relative cost	Low	Medium	High	High

The stringent vacuum requirements and the relatively poor stability mean that the cold field emitter is rarely used for Auger electron spectroscopy. For medium- and high-performance Auger instruments, either LaB_6 or Schottky field emitters are generally used in commercially available instruments.

The spot size attainable with a particular electron gun is a function of the primary beam current. For example, the smallest spot size obtainable on a scanning Auger microscope with electromagnetic lenses and LaB_6 filament is about 20 nm at 0.1 nA but increases to 100 nm at 10 nA. The intensity of Auger electrons emitted depends on the specimen current and at 0.1 nA spectrum acquisition will be a lengthy process, but at 10 nA the current of Auger electrons will be increased to the point where analysis becomes practical. A satisfactory compromise must be reached between spatial resolution and spectral intensity. A field emission source in a similar column will give somewhat better resolution but at a much improved current owing to its superior brightness. Such electron guns are the preserve of high-resolution scanning Auger microscopes; such a gun is illustrated in Figure 2.6.

When SAM is required on a multi-technique XPS instrument, the best configuration probably includes a high-brightness, hot field emission

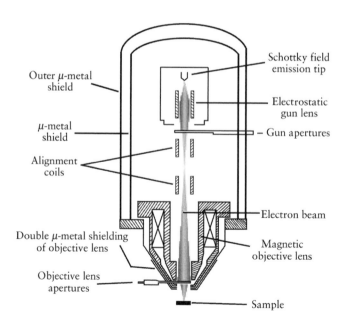

Figure 2.6   Schematic diagram of a UHV electron gun for Auger electron spectroscopy

source with electrostatic lenses. Such an assembly will provide reliable routine operation with the minimum of attention and an analytical spot size of < 100 nm.

2.5 Analysers for Electron Spectroscopy

There are two types of electron energy analyser in general use for XPS and AES – the cylindrical mirror analyser (CMA) and the hemispherical sector analyser (HSA). The CMA is used when it is not important that the highest resolution is achieved and when it is necessary to collect electrons from only a small area (< 1 mm diameter); for example, it can be used for AES if chemical state information is not required. Typically, commercial CMAs can provide an energy resolution of about 0.4 to 0.6 per cent of the energy to which they are tuned. HSAs typically achieve more than a factor of ten better than this.

The development of these two analysers is a reflection of the requirements of AES and XPS at the inception of the techniques. The primary requirement for Auger spectroscopy was that of high sensitivity (analyser transmission), the intrinsic resolution (the contribution of analyser broadening to the resultant spectrum) being of minor importance. The need for high sensitivity led to the development of the CMA. For XPS, on the other hand, it is spectral resolution that is the cornerstone of the technique and this led to the development of the HSA as a design of analyser with sufficiently good resolution. The addition of a transfer lens to the HSA and multi-channel detection increases its sensitivity to the point where both high transmission and high resolution are possible and this type of analyser may now be used with excellent results for both XPS and Auger electron spectroscopy. The nature of the CMA does not permit the addition of a transfer lens.

2.5.1 The cylindrical mirror analyser

The CMA consists of two concentric cylinders as illustrated in Figure 2.7, the inner cylinder is held at earth potential while the outer is ramped at a negative potential. An electron gun is often mounted

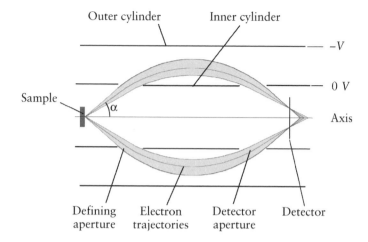

Figure 2.7 Schematic diagram of the cylindrical mirror analyser (CMA)

coaxially within the analyser. A certain proportion of the Auger electrons emitted will pass through the defining aperture in the inner cylinder and, depending on the potential applied to the outer cylinder, electrons of the desired energy will pass through the detector aperture and be re-focused at the electron detector. Thus, an energy spectrum – the direct energy

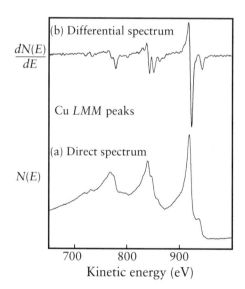

Figure 2.8 Comparison of (a) direct, and (b) differential Auger spectra for copper

spectrum – can be built up by merely scanning the potential on the outer cylinder to produce a spectrum of intensity (in counts per second) versus electron kinetic energy. This spectrum will contain not only Auger electrons but all the other emitted electrons, the Auger peaks being superimposed, as weak features, on an intense background.

For this reason the differential spectrum is often recorded rather than the direct energy spectrum. This used to be achieved by applying a small a.c. modulation to the analyser and comparing the output at the detector with this standard a.c. signal by means of a phase-sensitive detector (lock-in-amplifier). The resultant signal is displayed as the differential spectrum. In modern Auger spectrometers, however, the differential spectrum, if required, is calculated within the data system. A comparison of direct (pulse counted) and differential Auger electron spectra, from a copper foil, is presented in Figure 2.8.

While the CMA can provide good sensitivity for AES, it suffers from a number of disadvantages which make it totally unsuitable for XPS.

- The optimum resolution obtainable with this type of analyser is not good enough to provide the chemical state information which is available from XPS.

- The energy calibration of the analyser is dependent upon the position of the sample surface along the axis of the analyser.

- The area from which electrons can be collected is very small.

In an attempt to overcome these disadvantages, a double pass CMA was developed with limited success but all modern, commercial XPS instruments are now equipped with an HSA.

2.5.2 The hemispherical sector analyser

A hemispherical sector analyser (HSA), otherwise known as a concentric hemispherical analyser (CHA) or a spherical sector analyser (SSA), consists of a pair of concentric hemispherical electrodes between which there is a gap for the electrons to pass. Between the sample and the analyser there is usually a lens, or a series of lenses.

The lenses serve a number of purposes which will be discussed later but it is helpful to mention one now. The kinetic energy of the electrons as they are ejected from the sample is usually too great for the analyser to produce sufficiently high resolution so they must be retarded. This retardation is achieved either within the lens or, using parallel grids, between the lens and the analyser.

The schematic diagram of Figure 2.9 shows a typical HSA configuration for XPS. A potential difference is applied across the two hemispheres with the outer hemisphere being more negative than the inner one. Electrons injected tangentially at the input to the analyser will only reach the detector if their energy is given by

$$E = e\Delta V \left(\frac{R_1 R_2}{R_2^2 - R_1^2} \right)$$

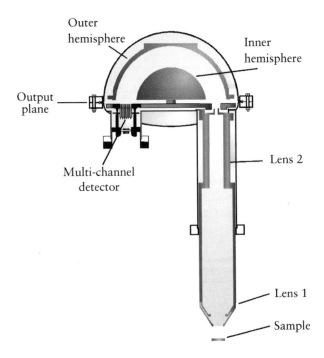

Figure 2.9 Schematic diagram of a modern HSA and transfer lens

where the kinetic energy of the electrons is given by E, e is the charge on the electron, ΔV is the potential difference between the hemispheres and R_1 and R_2 are the radii of the inner and outer hemispheres respectively. The radii of the hemispheres is constant and so the above equation can be expressed as

$$E = ke\Delta V$$

in which k is known as the spectrometer constant and depends upon the design of the analyser.

An HSA also acts as a lens and so electrons entering the analyser on the mean radius will reach the exit slit even if they enter the analyser at some angle with respect to the tangent to the sphere given by the mean radius.

Electrons whose energy is higher than that given by the above expression will follow a path whose radius is larger than the mean radius of the analyser and those with a lower kinetic energy will follow a path with smaller radius. Provided that the energy of these electrons does not differ too greatly from that given by the expression, these electrons will also reach the output plane of the analyser. It is possible therefore to provide a number of detectors at the output plane. These detectors are arranged radially across the output plane. Clearly, each of the detectors collects electrons of a different energy but by adding the signal into the appropriate energy channel the sensitivity of the instrument can be increased by a factor equal to the number of detectors, if the size of the detector does not change. Instruments with up to nine discrete channel electron multipliers (channeltrons) are commercially available. Some instruments are fitted with two-dimensional detectors which allow the user to select the number of channels collected at any one time. Up to 112 channels can be defined on some instruments, the width of each channel is small so this does not imply that the sensitivity is 112 times that of an equivalent instrument fitted with a single channeltron. The advantage of having such a large number of channels is that it allows high-quality spectra to be recorded without scanning the analyser.

The HSA is typically operated in one of two modes: constant analyser energy (CAE), sometimes known as fixed analyser transmission (FAT), and constant retard ratio (CRR) also known as fixed retard ratio (FRR).

CAE mode of operation

In the CAE mode electrons are accelerated or retarded to some user-defined energy which is the energy the electrons possess as they pass through the analyser (the pass energy). In order to achieve analysis in the CAE mode, the voltages on the hemispheres are scanned according to the graph shown in Figure 2.10.

The selected pass energy affects both the transmission of the analyser and its resolution. Selecting a small pass energy will result in high resolution while a large pass energy will provide higher transmission but poorer resolution. The pass energy remains constant throughout the energy range thus the resolution (in eV) is constant across the entire width of the spectrum.

The range of pass energies available to the user depends upon the design of the spectrometer but is typically from about 1 eV to several hundred electron volts. Figure 2.11 shows part of the XPS spectrum of silver recorded at a series of pass energies, showing the effect of pass energy on resolution and sensitivity. In a typical XPS experiment, the user will select a pass energy in the region of 100 eV for survey or wide scans and in the region of 20 eV for higher-resolution spectra of individual core levels. These narrow scans are used to establish the chemical states of the elements present and for quantification purposes. It is normal practice to collect XPS spectra in the CAE mode.

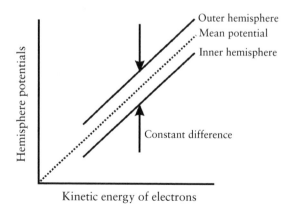

Figure 2.10   Operation of the HSA in constant analyser energy mode

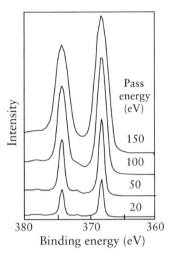

Figure 2.11 XPS of Ag showing the effect of pass energy upon the Ag 3d part of the spectrum

CRR mode of operation

In the CRR mode electrons are retarded to some user-defined fraction of their original kinetic energy as they pass through the analyser (the retard ratio). In order to achieve analysis in the CRR mode, the voltages on the hemispheres are scanned according to the graph shown in Figure 2.12.

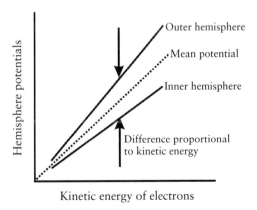

Figure 2.12 Operation of the HSA in constant retard ratio mode

In this mode the pass energy is proportional to the kinetic energy:

$$\text{Pass energy} = \frac{\text{Kinetic energy}}{\text{Retard ratio}}$$

The percentage resolution in this mode of operation is constant and inversely proportional to the retard ratio. The constant of proportionality depends upon the design of the instrument, on a typical instrument

$$\text{Resolution} \approx (2/\text{Retard ratio})\%$$

Typically, the resolution available from a good commercial HSA can be selected from the range 0.02 per cent to 2.0 per cent, in the case of the typical spectrometer referred to above, this means that the range of retard ratios is from 1 to 100.

It is normal practice to collect Auger electron spectra in the CRR mode. When collecting a survey spectrum or when only elemental information is required, a resolution in the region of 0.5 per cent is often considered to provide a good compromise between sensitivity and resolution. If chemical state information is needed then the resolution has to be better than this and a value in the range 0.02 per cent to 0.1

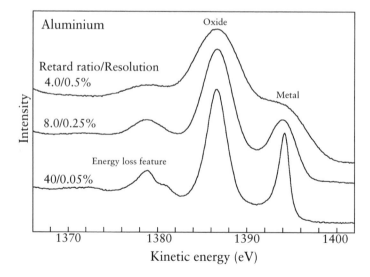

Figure 2.13 Auger spectra of aluminium and its native oxide showing the effect of changes in the retard ratio upon the resolution of the spectrum; at high retard ratio the metal, oxide and energy loss features become resolved

per cent is usually chosen depending upon the size of the chemical shift that needs to be observed.

Figure 2.13 shows the affect of changing the retard ratio on the Auger spectrum of aluminium metal which is covered with a thin layer of oxide. If differential spectra are needed these are usually calculated from the direct spectra within the data system, rather than using phase sensitive detection which was popular in the past.

Comparison of CAE and CRR modes

Figure 2.14 shows the XPS spectrum of copper collected in CRR and CAE modes. The retard ratio in Figure 2.14(a) is 4 and the pass energy in Figure 2.14(b) is 100 eV. This means that the transmission and

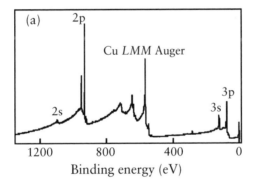

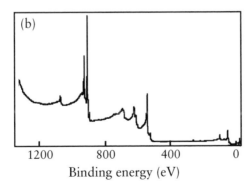

Figure 2.14 XPS survey spectra from copper acquired in (a) CRR mode (retard ratio = 4), and (b) CAE mode (pass energy = 100 eV)

resolution are identical at a kinetic energy of

$$4 \times 100 = 400\,\text{eV},$$

that is at a binding energy of $1087\,\text{eV}$, the approximate position of the Cu 2s peak.

In the CRR mode the pass energy is proportional to the kinetic energy and so the resolution, ΔE, becomes worse with increasing kinetic energy (decreasing binding energy) but the relative resolution, expressed as a percentage of the kinetic energy, $\Delta E/E$, is constant throughout the energy range. The transmission increases with increasing kinetic energy, which has the effect of suppressing the relatively high electron yield at the low kinetic energy end of the spectrum. This makes this mode of operation ideal for analysis using Auger electron spectroscopy. As both spectral resolution and transmission change with electron energy, quantification of XPS spectra is difficult in the CRR mode of operation and CAE is generally preferred.

In the CAE mode of operation, the resolution, expressed in eV, and the transmission of the analyser both remain constant throughout the energy range of the scan. This ensures that XPS quantification is more reliable and accentuates the XPS peaks at the low kinetic energy (high binding energy) end of the spectrum.

The transfer lens

The performance of the HSA is strongly dependent upon the nature and quality of the transfer lens or lenses between the sample and the entrance to the analyser. The presence of a transfer lens:

- moves the analyser away from the analysis position allowing other components of the spectrometer to be placed closer to the sample,

- maximizes the collection angle to ensure high transmission and sensitivity,

- retards the electrons prior to their injection into the analyser,

- determines and controls the area of the sample from which electrons are collected, allowing small area XPS measurements to be made,

- controls the acceptance angle – this has obvious applications in defining the angular resolution for angle resolved XPS, and it is also important for small area and imaging XPS because the angular acceptance will determine the spatial resolution as well as the transmission.

In the past, these transfer lenses were exclusively electrostatic but there are some modern instruments which are fitted with a magnetic immersion lens. This is a large electromagnet the current through which is scanned as a function of the kinetic energy of the electrons being analysed. When using this type of lens, the specimen is situated within the magnetic field. A major advantage of this type of lens is that it can collect electrons from a very wide range of emission angles; typically, a collection angle of 90° can be achieved compared with about 25°, usual for a purely electrostatic system. This can significantly improve the sensitivity of the instrument. Magnetic lenses can provide better spatial resolution than electrostatic lenses of the same focal length because their aberration coefficients are lower.

Magnetic lenses cannot be used for AES because the magnetic field would deflect the primary electron beam and affect the spot size at the sample surface. The extent of these effects would depend upon the magnetic field strength, which is varied with the kinetic energy being analysed.

2.6 Detectors

In most electron spectrometers it is necessary to count the individual electrons arriving at the detector. To achieve this, electron multipliers are used. Although there are many types of electron multiplier, only two types are commonly used in electron spectrometers: channel electron multipliers (channeltrons) and channel plates.

2.6.1 Channel electron multipliers

These consist of a spiral-shaped glass tube with a conical collector at one end and a metal anode at the other. The internal walls of the detector are coated with a material which, when struck by an electron

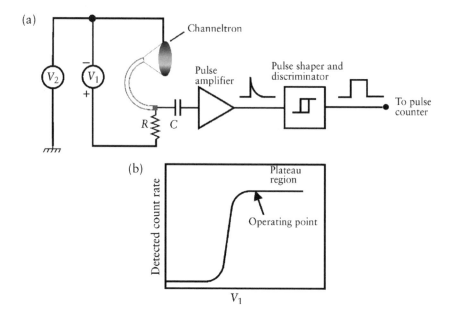

Figure 2.15 Schematic diagram showing (a) the operation of a channel electron multiplier, and (b) the output characteristics

having more than some threshold kinetic energy, will emit many secondary electrons. A large potential difference is applied across the length of the channeltron, the cone being negative (Figure 2.15).

As an electron strikes the internal surface of the cone, many electrons are emitted and accelerated into the tube of the detector where more collisions take place with the total number of electrons in the cascade increasing with each collision. The gain of a channeltron depends upon the potential difference between its ends. When the voltage is low no output pulses are detected. As the voltage increases above some threshold, pulses can be detected. The detection efficiency continues to increase with voltage until a plateau voltage is reached. Above this voltage, the measured output count rate is independent of the voltage across the channeltron. The operating point should be chosen to be just above the plateau. The precise operating voltage will depend upon the age of the channeltron and its design. Typically it will be between 2 kV and 4 kV. It is necessary to check periodically that the voltage across the channeltron is correct.

Each electron arriving at the detector typically results in about 10^8 electrons reaching the anode. It is necessary to amplify these pulses of charge using a pulse amplifier and produce a square wave which can be counted by a rate meter. Use of a discriminator eliminates noise signals emanating from the channeltron or the preamplifier.

The channeltron samples about 5 mm in the dispersive direction of the analyser and about 15 mm in the non-dispersive direction. In order to increase the sensitivity of the spectrometer it is common to use an array of channeltrons as the detector. The channeltrons are arranged along the dispersive direction and so each one collects a different electron kinetic energy. The data system sums the output from each channeltron after applying the appropriate energy shift. Channeltrons can be capable of detecting up to about 3×10^6 counts/s although they become non-linear at very high count rates.

2.6.2 Channel plates

A channel plate is a disc having an array of small holes. Each of these holes behaves as a small channeltron. The gain of an individual channel is much lower than that of a channeltron so it is common to use a pair of channel plates in tandem. The maximum count rate which can be detected using channel plates is about 3×10^5 counts/s on current systems for two-dimensional detection, but in principle could be as high as 1×10^7 counts/s for simple spectroscopy.

Channel plates are used when it is necessary to detect data in two dimensions; spectrometers have been designed using channel plates to measure signals:

- in an $X-Y$ array for parallel acquisition of photoelectron images,

- in an X–energy array for parallel acquisition of XPS line scans,

- in an energy–angle array for the parallel acquisition of angle resolved XPS spectra.

2.7 Small Area XPS

It is often desirable to analyse a small feature or imperfection on the surface of a sample. For the analysis to be effective, as much as possible of the signal from the surrounding area should be excluded. This is usually done in one of two ways.

1. By flooding the analysis area with X-rays but limiting the area from which the photoelectrons are collected using the transfer lens. This is described as lens-defined small area analysis.

2. By focusing a monochromated beam of X-rays into a small spot on the sample, described as source-defined small area analysis.

2.7.1 Lens-defined small area XPS

In most spectrometers the transfer lens fitted to the analyser is operated in such a way as to produce a photoelectron image at some point in the electron optical column. If a small aperture is placed at this point then only electrons emitted from a defined area of the sample can pass through the aperture and reach the analyser. If the magnification of the lens is M and the diameter of the aperture is d then the diameter of the analysed area is d/M. In some instruments, an aperture can be selected from a number of fixed apertures while in other instruments an iris is used to provide a continuous range of analysis areas.

Spherical aberrations, which occur in any electron optical lens system, mean that the acceptance angle of the lens has to be limited to provide good edge resolution in the analysis area. This is achieved using either another set of fixed apertures or another iris placed at some point remote from the image position of the lens. Using this technique, commercial instruments can provide small area analysis down to about 15 μm.

This is an effective method for producing high-quality, small area XPS data but it suffers from a disadvantage. Reducing the angular acceptance of the lens reduces the detected flux per unit area of the sample. This means that analysis times can become very long. During the analysis time the whole of the sample (or samples in a multi-sample experiment) is being exposed to X-rays, potentially resulting in radiation damage. Therefore, if many samples or many points on a single sample are to be analysed then the analyst cannot be certain that the surface remains unaltered.

2.7.2 Source-defined small area analysis

It is possible to shape a quartz crystal so that it can focus a beam of electrons *and* provide monochromatic X-rays by diffraction. In this respect, it behaves rather like a concave mirror. The focusing is usually achieved using a magnification of unity which means that the size of the X-ray spot on the sample is approximately equal to the size of the electron spot on the X-ray anode. Analysis areas down to about 10 μm can be achieved in commercial instruments using this method.

Because the source of X-rays is defining the analysis area, aberrations in the transfer lens will not affect the analysis area and so the lens can be operated at its maximum transmission, regardless of how small the analysis area becomes. The sensitivity of a spectrometer operating in this mode is therefore much higher than that of an equivalent instrument operating in the lens defined mode. This reduces the danger of sample damage during analysis and eliminates radiation damage to the surrounding area of the sample(s).

The intensity of the X-ray beam is proportional to the intensity of the electron beam at the anode. The intensity of the electron beam at the anode is limited by the rate at which heat can be removed from the vicinity of the electron spot. The X-ray source can be operated at several hundred watts for large area analysis while it is only possible to use a few watts at the smallest areas.

2.8 XPS Imaging and Mapping

A logical extension to small area XPS is to produce an image or map of the surface. Such an image or map shows the distribution of an element or a chemical state on the surface of the sample. There are two distinct approaches, used by manufacturers, to obtaining XPS maps.

1. Serial acquisition in which each pixel of the image is collected in turn.

2. Parallel acquisition in which data from the whole of the analysis area is collected simultaneously.

2.8.1 Serial acquisition

Serial acquisition of images is based on a two-dimensional, rectangular array of small area XPS analyses. By this means, the distribution of elements or chemical states can be measured. The ultimate spatial resolution in the image is determined by the size of the smallest analysis area (this depends upon the instrument but 10 μm is possible using a high-quality, modern source-defined spectrometer). Serial acquisition is generally slower than parallel acquisition but has the advantage that one can collect a range of energies at each pixel whereas, in parallel acquisition, only a single energy can be collected during each acquisition.

There are several methods by which the analysis area can be stepped over the field of view of the image.

1. *Scanning the sample stage.* Using this method, the analysis position is fixed in space and the sample surface is moved with respect to this position. The advantages of this method are that all of the important instrumental conditions remain constant (e.g., the energy of the X-ray beam, the resolution of the analysis spot, and the transmission function of the lens) and the maximum size of the image field of view is limited only by the range of motion of the sample stage. The disadvantages of the method are that it tends to be slower than other methods and requires a high-precision stage with low backlash.

2. *Scanning the lens.* This method requires that two pairs of deflector plates be built into the transfer lens. By applying potentials to these plates, the photoelectron image can be deflected with respect to the area-defining aperture within the transfer lens. The analysis area can therefore be scanned in the X and Y directions and a map built up. The advantage of this method is that it is faster than scanning the sample stage but it suffers from a major disadvantage. The resolution of the map rapidly degrades as a function of distance from the centre of the map due to the spherical aberrations which are inevitably present in the electrostatic lens.

3. *Scanning the monochromated X-ray spot.* As mentioned in the discussion of small area XPS, it is possible to produce an X-ray image

of the spot of electrons on the X-ray anode. Therefore, if the spot of electrons is scanned on the anode, the X-ray beam will be scanned on the sample surface. Again, the transfer lens can be operated in its maximum transmission mode because it does not contribute to the spatial resolution. The advantages of this method are the same as those for source-defined small area XPS. The disadvantage is that the field of view is very limited. This is especially true in the direction of the Bragg angle. As the electron beam is scanned over the anode, the diffraction angle is changing and so the wavelength of the X-rays reaching the sample is changing. In turn, this means that the kinetic energy of the photoelectrons will change and so the energy to which the analyser is tuned must be adjusted as a function of distance. If the deflection angle is too large there will be no intensity in the AlKα radiation at the required wavelength and so photoemission is not possible. If it is necessary to map large areas of the sample, this method must be used in combination with stage scanning.

2.8.2 Parallel acquisition

In parallel acquisition of photoelectron images, the whole of the field of view is imaged simultaneously without scanning voltages being applied to any component of the spectrometer.

To obtain images via this route, additional lenses are required in the spectrometer. These must also be equipped with a two-dimensional detector. Figure 2.16 shows schematically how the method works. The photoelectrons pass through lenses 1 and 2 in the transfer lens assembly, producing a photoelectron image of the specimen surface at some plane within the lens column after each lens (the image planes). Lens 3 is operated such that its focal length is equal to the distance between the lens and the second image plane. This means that electrons emanating from any one point on the image will leave lens 3 on parallel paths. The angle between the beam of electrons and the lens axis will depend only upon their distance from the lens axis in the image plane. The electrons then enter the analyser, which functions as both an energy filter *and* a lens. If the deflection angle of the analyser is 180° then the angular distribution of the electrons as they leave the analyser is

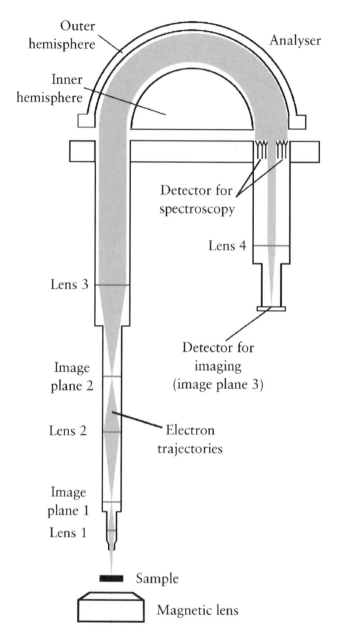

Figure 2.16 Schematic diagram illustrating the principles of parallel acquisition of photoelectron images

retained. Therefore a fourth lens (lens 4) operated in the reverse manner to that of lens 3 will reconstruct the image at a two-dimensional detector placed at the focal length of lens 4.

The spatial resolution of parallel imaging is dependent upon the spherical aberrations in the lens. Limiting the angular acceptance of the lens can reduce the effect of the aberrations and so resolution can be improved at the expense of sensitivity. The use of a magnetic immersion lens in the sample region also reduces aberrations and therefore allows higher sensitivity at a given resolution. This method of imaging is relatively fast and commercial instruments can produce images with an image resolution of $<3\,\mu m$.

Parallel imaging clearly provides the best image resolution and is faster than the serial methods but it only collects an image at a single energy. It is customary to make a measurement at a photoelectron peak energy and a second measurement at some energy remote from the peak,

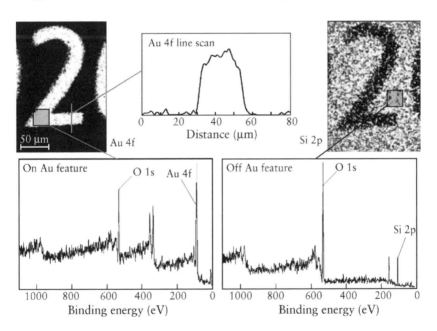

Figure 2.17 Examples of parallel XPS images from gold features on a glass substrate; the resolution in the Au 4f and Si 2p images can be measured from the line scan to be about 3 μm, and the images can be used to define small areas from which spectra can be acquired

where the signal intensity is approximately equal to the estimated background signal under the peak maximum. By subtracting the background signal from the signal at the peak maximum, a more accurate measurement can be made. This is in contrast with the use of a serial mapping method in conjunction with a multi-channel detector, to produce a 'snapshot' spectrum at each pixel of the map. Such spectra can then be treated with advanced data processing techniques in order to extract the maximum chemical information from the image.

Figure 2.17 shows examples of XPS images from parallel acquisition using monochromated AlKα X-rays. This sample was a glass substrate with gold features. Images from Au 4f and Si 2p are shown along with an intensity line scan measured from the Au 4f image. The line scan indicates that the lateral resolution in this image is about 3 µm. These images can be used to define areas for spectroscopy. A 25 µm square area was selected from the gold region of the sample and from the glass – the respective spectra are also shown in Figure 2.17. Images such as these can be acquired in just a few minutes, those in Figure 2.17 were collected in 4 min (2 min for the peak image and 2 min for the background) and the spectra were collected in less than 2 min each. The sample was an insulator and so required the use of an electron flood gun to control the sample charging.

2.9 Lateral Resolution in Small Area XPS

To estimate the lateral resolution in small area XPS, the following practice is adopted. Assume, initially, that the analysis area in a small area mode of acquisition is circular and, outside this area, there is no transmission. Transmission is uniform within the area (i.e., the transmission as a function of position on the sample resembles a 'top hat' distribution). To determine the dimensions of the analysis area a knife-edge sample (often silver) is translated through the analysis area while measuring the XPS signal from the knife-edge (e.g., the signal from the Ag $3d_{5/3}$ peak), see Figure 2.18. The signal is initially zero until the knife-edge begins to intercept the analysis area when it begins to rise. The intensity of the signal continues to rise until the silver completely fills the whole of the analysis area.

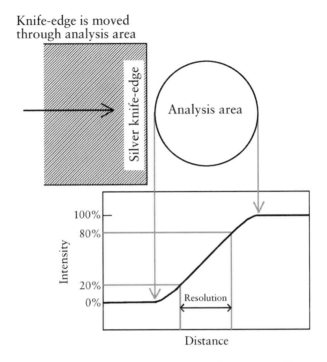

Figure 2.18   Measurement of the spatial resolution in small area XPS

The signal is then plotted against distance, as shown in Figure 2.18. The distance through which the knife-edge has to be translated for the signal to change between two prescribed percentages of the total signal change is then determined. This distance is then described as the lateral resolution for spectroscopy. The percentages used depend upon the instrument manufacturer. The range 20 per cent to 80 per cent is used by some because the reported spatial resolution is then approximately equal to the radius of the analysis area (within <2 per cent). If the response of the spectrometer can be accurately represented by a Gaussian curve, instead of a top hat, then a value of 16 per cent to 84 per cent represents the distance between the two points each one standard deviation on either side of the centre of the analysis area.

The method described above is specifically for the case when the analysis area is defined by the transfer lens ahead of the analyser (lens-defined small area XPS). The same method can be used when the area is defined by the X-ray spot size (source-defined small area XPS).

Measurement of lateral resolution should be made in two orthogonal directions in case the analysis area is not circular.

2.10 Angle Resolved XPS

As mentioned in the previous chapter, the finite mean free path of electrons within a solid means that the information depth in XPS analysis is of the order of a few nanometers. This, of course, is only true if the electrons are detected at a direction normal to the sample surface. If electrons are detected at some angle to the normal, the information depth is reduced by an amount equal to the cosine of the angle between the surface normal and the analysis direction. This is the basis for a powerful analysis technique, angle resolved XPS (ARXPS). One of the reasons for the usefulness of the method is that it can be applied to films which are too thin to be analysed by conventional depth profiling techniques or those that are irretrievably damaged by such methods (e.g., polymers). Another reason to use ARXPS is that it is a non-destructive technique which can provide chemical state information, unlike methods based upon sputtering.

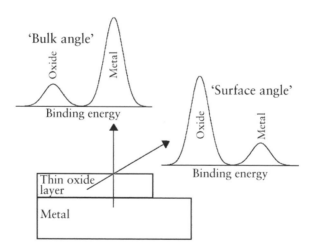

Figure 2.19 Illustration of XPS spectra taken from a thin oxide film on a metal at near normal collection angle (bulk angle) and near grazing collection angle (surface angle)

To obtain ARXPS data the angular acceptance of the transfer lens is set to provide good angular resolution, usually the half angle is set to be in the region of 1° to 3°. A series of spectra is then acquired as the sample surface is tilted with respect to the lens axis. Figure 2.19 illustrates how the spectra might appear at each end of the angular range if the sample consists of a thin oxide layer on a metal substrate. Note that the relative intensity of the oxide peak is larger at the near grazing emission angle.

ARXPS measurements such as this can provide information about the thickness and chemical composition of thin surface layers, as will be illustrated in a later chapter.

There is one commercial instrument which produces ARXPS spectra without tilting the sample; it is capable of parallel collection of angle resolved data. The two-dimensional detector at the output plane has photoelectron energy dispersed in one direction (as with a conventional lens analyser arrangement) and the angular distribution dispersed in the other direction, (Figure 2.20). Such an arrangement can provide an angular range of 60° with a resolution close to 1°. This has a number of advantages over the conventional method.

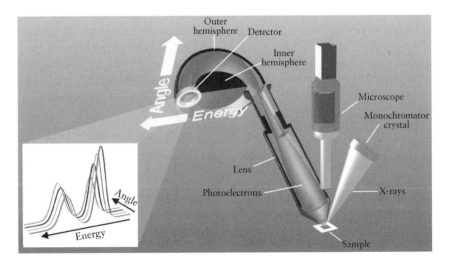

Figure 2.20 The arrangement of a spectrometer capable of collecting angle resolved spectra in parallel

- ARXPS measurements can be taken from very large samples, such as complete semiconductor wafers. Such samples are too large to be tilted inside an XPS spectrometer.

- The analysis position remains constant throughout the angular range. When combining small area XPS and ARXPS it is difficult to ensure that the analysis point remains fixed during the experiment especially if that point is remote from the tilt axis even with a eucentric stage. Since the sample does not move during parallel collection, the analysis position remains constant.

- The analysis area remains constant during the analysis. If lens-defined small area XPS is combined with ARXPS then the analysis area would increase by a large factor as the sample is tilted away from its normal position. Using a combination of source-defined small area analysis and parallel collection the analysis area becomes independent of angle.

The use of ARXPS as a non-destructive method for near-surface depth profiling of samples will be discussed further in Chapter 4.

3 The Electron Spectrum: Qualitative and Quantitative Interpretation

The product of the electron spectrometer is amenable to many levels of interpretation, ranging from a simple qualitative assessment of the elements present to a full-blown quantitative analysis complete with assignments of chemical states, and determination of the phase distribution for each element. In practice, a happy medium is usually required with an estimation made of the relative amounts of each element present. There are certain similarities in the way that AES and XPS spectra are treated. We shall *initially* consider them together as this also provides a means of comparing the analytical capabilities of the two methods.

3.1 Qualitative Analysis

The first step to be taken in characterizing the surface chemistry of the specimen under investigation is the identification of the elements present. To achieve this it is usual to record a survey, or wide scan, spectrum over a region that will provide fairly strong peaks for all elements in the periodic table. In the case of *both* XPS *and* AES, a range of 0–1000 eV is often sufficient. The current IUVSTA[4] recommendations

---

[4]IUVSTA is the International Union for Vacuum Science, Technique and Applications.

Table 3.1 IUVSTA recommended conditions for
the acquisition of a survey spectrum

	$MgK\alpha$	$AlK\alpha$
Energy range (eV)	0–1150	0–1350
Energy step size (eV)	0.4	0.4

for the acquisition of XPS survey spectra extend this range, as shown in
Table 3.1.

The individual peaks may be identified with the aid of data in tabular or
graphical form as reproduced in Appendices 1 and 2. A typical differential Auger spectrum is shown in Figure 3.1(b). The peaks produced by the
elements present, in this case Al, O, and C, are observed superimposed on
a background typical of Auger spectra. Auger spectra may be recorded in
either the direct (Figure 3.1(a)), or the differential (Figure 3.1(b)) mode.
Nowadays, the direct mode is rather more popular with the advent of
high spatial resolution scanning Auger microscopes. The photoelectron
spectrum from a similar specimen, Figure 3.1(c) is composed of the individual photoelectron peaks and the associated Auger lines resulting
from the de-excitation process following photoemission. Unlike the
Auger spectrum of Figure 3.1(b) the electron background is relatively
small and increases in a step-like manner after each spectral feature. This
is a result of the scattering of the characteristic Auger or photoelectrons
within the matrix, bringing about a loss of kinetic energy. The shape of
this background itself contains valuable information and, to the experienced electron spectroscopist, provides a means of assessing the way in
which near-surface layers are arranged. In the case of a perfectly clean
surface, the photoelectron peaks will have a horizontal background or
one with a slightly negative slope; if the surface is covered with a thin
overlayer the peaks from the buried phase will have a positive slope, in the
most severe case the peak itself will be absent and the only indication will
be a change in background slope at the appropriate energy.

3.1.1 Unwanted features in electron spectra

The XPS spectrum is further complicated by the presence of several
features of no analytical use such as X-ray satellites and X-ray ghosts.

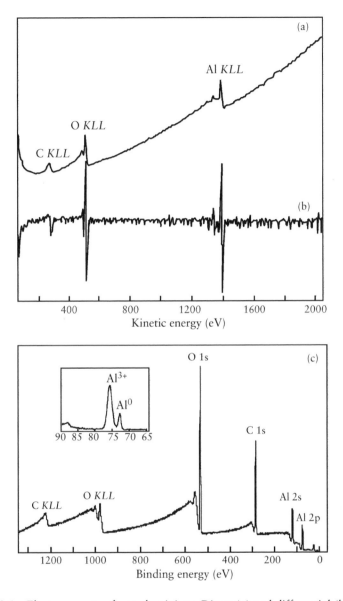

Figure 3.1 Electron spectra from aluminium: Direct (a) and differential (b) Auger spectra, and (c) the XPS survey spectrum (recorded with monochromatic Al Kα radiation) of oxidized aluminium foil following air exposure; note the large C 1s peak resulting from the deposition of adventitious carbon from the atmosphere; inset is the Al 2p region recorded at higher resolution showing the metallic (Al^0) and oxide (Al^{3+}) components

X-ray satellites are present if unmonochromated radiation is used, and occur because the characteristic transitions are excited by a minor component of the X-ray spectrum, e.g., $AlK\alpha_{3,4}$, $AlK\alpha_{5,6}$, $AlK\beta$. Such features are *to be expected* and only present difficulties if they occur at the same binding energy as an element present in very small concentrations. The solution may then be to change to another radiation ($MgK\alpha$) as the separations are slightly different. The $AlK\alpha_{3,4}$, X-ray satellite is easily identified in the XPS spectrum of Figure 2.14 (page 43) for the most intense photoelectron peak (Cu2p), as a small peak at a binding energy of approximately 920 eV. Auger transitions present in an XPS spectrum do not show such satellite features and this provides a rapid means of distinguishing between the two as seen on the O $KL_{2,3}L_{2,3}$ peak of Figure 3.1.

X-ray ghosts arise from unsuspected X-rays irradiating the sample; these may result from 'crosstalk' in a twin anode gun (the generation of a small amount of characteristic X-radiation as a result of anode misalignment), from the anode material not in use in addition to the X-ray flux from the chosen source, or possibly $CuL\alpha$ ($h\nu = 929.7$ eV) from the exposed base material of a damaged anode. In each case, the problem should be reduced to an inconsequential level by overhauling and readjusting the X-ray gun.

3.1.2 Data acquisition

The survey spectrum in XPS or AES will generally be followed up by the acquisition of spectra around the elemental peaks of interest. As both XPS and AES spectra contain valuable chemical state information these regions will be recorded at a higher resolution. At this point it is appropriate to consider the factors that influence the physical width of the Auger or photoelectron features. As the Auger process is a de-excitation process involving three electrons, any uncertainty in the energies of the electrons involved will be observed in the resultant spectrum. Thus, Auger transitions involving electrons from the band structure of the atom will be broad. Indeed for the best quality of chemical state information in AES the transitions of choice are those involving three core electrons (CCC transitions). This said, excellent chemical state information can be obtained from *CVV* Auger electrons such as the Cu L_3VV. Thus, the natural width of Auger transitions is dominated by the process itself rather than instrumental factors.

In XPS, the natural line width is very small, although it can influence the shape of the spectrum when recorded at very high resolution. To obtain high-quality core level spectra, a monochromated AlKα source is used which provides an X-ray line width of 0.25 eV in the best cases (cf. 0.7 for MgKα and 0.9 eV for AlKα, from achromatic sources). The narrowness of the monochromated source is combined with a very high-resolution mode for the concentric hemispherical analyser (CAE mode at, typically, 5 or 10 eV pass energy). This enables fine structure in the core level spectra to be seen and the rapid increase in the use of monochromatic sources in the early 1990s led to a step change in the level of information attainable from core level spectra in XPS, particularly for the analysis of polymers. The complexity of core level XPS spectra recorded at high resolution has seen a parallel growth in the level of sophistication of peak fitting routines to enable the analyst to assign the various components of a convoluted spectrum with a degree of confidence. Important parameters in such analysis include: the number of components, the shape of the peak (usually a Voight function), degree of asymmetry (important in metals which are asymmetric as a result of core hole lifetime effects), the width of the components (constrained together or allowed to vary), and the shape of the background which will be influenced by both elastic and inelastic scattering of the electrons. The source and use of chemical state information in both XPS and AES is discussed in detail in Section 3.2.

The above discussion relates to core level XPS peaks but important information is contained in the valence band region, which in practice extends from the Fermi level up to a binding energy of 30 eV. This region of the XPS spectrum is very weak, as can be seen from any survey spectrum, a problem that is overcome nowadays with the use of high intensity monochromatic sources and high transmission analysers. However, it has been used as a fingerprint region of the spectrum since the very early days of XPS. It has the advantage that specimens that give very similar, or identical, core level spectra can be distinguished by examination of the valence band. This is seen in the valence band regions for diamond, graphite and polyethylene – clear differences can be seen, unlike the C 1s spectra which are very similar. The restrictions on the use of X-rays for probing the valence band are the poor cross-sections and the line width of the source. Both can be overcome by turning to lower energy photons such as HeI, HeII, NeI or NeII (with photon energies of 21.2 eV, 40.8 eV, 16.8 eV, and 26.9 eV respectively) which provide line widths in the region of 100 meV. This form of photo-emission is known as ultraviolet photoelectron spectroscopy (UPS). Unlike XPS, this is a molecular spectroscopy which provides no direct elemental information.

3.2 Chemical State Information

3.2.1 X-ray photoelectron spectroscopy

Long before XPS had developed into a commercially available method for surface analysis it was clear that the spectra produced as a consequence of X-irradiation exhibited small changes in electron energy that were a result of the chemical environment of the emitting atom, ion or molecule. This led Kai Siegbahn to coin the name Electron Spectroscopy for Chemical Analysis (ESCA), a term which is no longer in favour as a formal description of the technique but one which is still widely used colloquially and in the model names of various commercial spectrometers. The XPS chemical shift is the cornerstone of the technique and the reason high-resolution analysers and accurate calibration of energy scales were seen in XPS long before it was considered a necessity in AES.

Almost all elements in the periodic table exhibit a chemical shift, which can vary from a fraction of an electron volt up to several eVs. Set alongside the line width of the X-ray sources used in XPS (0.25–0.9 eV) it is clear why some form of data processing is often required to extract the maximum level of information from a spectrum. The computer curve fitting of high-resolution XPS spectra is now a routine undertaking, as indicated above, and international standards are now being drafted to provide a unified framework within which such procedures can be undertaken.

The shifts observed in XPS have their origin in either initial-state or final-state effects. In the case of initial-state effects, it is the charge on the atom prior to photoemission that plays the major role in the determination of the magnitude of the chemical shift. For example, the C–O bond in an organic polymer is shifted 1.6 eV relative to the unfunctionalized (methylene) carbon, while C=O and O–C–O are both shifted by 2.9 eV. In essence, the more bonds with electronegative atoms that are in place, the greater the positive XPS chemical shift. This is illustrated in a striking manner for fluoro-carbon species, the C 1s chemical shifts being larger than those of carbon–oxygen compounds as fluorine is a more electronegative element. The C–F group is shifted by 2.9 eV whilst CF_2 and CF_3 functionalities are shifted by 5.9 eV and 7.7 eV respectively. Unfortunately, such examples of the chemical shift are unusually large and, in general, values of 1–3 eV are encountered. An example of the manner in which the peak fitting of a complex C 1s spectrum is achieved is shown in Figure 3.2. The sample is an organic molecule, the diglycidyl ether of bisphenol A, which is a precursor to many thermosetting paints, adhesives and matrices for composite materials. By the consideration of the structure of the molecule it is possible to build up a synthesized spectrum, the relative intensities of the individual components reflecting the stoichiometry of the sample. For polymer XPS at this level of sophistication the resolution attainable with a monochromatic source is absolutely essential.

Final-state effects that occur following photoelectron emission, such as core hole screening, relaxation of electron orbitals and the polarization of surrounding ions are often dominant in influencing the magnitude of the

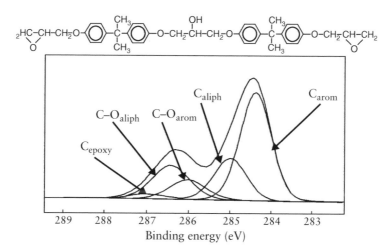

Figure 3.2 C 1s spectrum of the basic building block of epoxy product, the diglycidyl ether of bisphenol A, the structure of which is shown above the spectrum; this spectrum was recorded using monochromatic Al radiation

chemical shift. In most metals there is a positive shift between the elemental form and mono-, di- or trivalent ions but in the case of cerium the very large final-state effects give rise to a negative chemical shift of about 2 eV between Ce and CeO_2. This is, however, the exception and most elements behave in a predictable manner.

There are various compilations of binding energies and the most extensive is that promulgated by the National Institute of Standards and Technology, USA (NIST) which is available free of charge over the internet (http://srdata.nist.gov/xps/index.htm). This provides a ready source of standard data with which the individual components of a spectrum can be assigned with a high degree of confidence.

3.2.2 Electron induced Auger electron spectroscopy

Historically, electron induced AES is not credited with the ability to yield chemical state information. Early examples of chemical effects in Auger spectroscopy were usually in quasi-atomic spectra excited by X-rays. The reason for this neglect of chemical effects is twofold. The thrust in the early development of AES was the use of analysers, such as the retarding field or the cylindrical mirror analyser, which provided a high level of transmission but at the expense of spectral resolution. Thus, the peaks from early Auger spectrometers were very broad and superimposed on a very intense electron background. This led to the practice of using phase-sensitive detection to acquire differential spectra. Even if there were well-defined, chemical information in the spectra, the practices used for spectral acquisition would have effectively obliterated it! The other reason is the superposition of the degenerate band structure onto the shape of the Auger peak in the case of *VVV* and *CVV* transitions. This may lead to changes in shape of Auger transitions from different chemical environments but, generally, not the discrete chemical shift that is observed in XPS core levels. If the two outer electrons are not valence electrons (i.e., *CCC* Auger transitions) a sharp peak may result as observed for example in the *KLL* series of peaks of aluminium and silicon, and the *LMM* series of copper, zinc, gallium, germanium, and arsenic. The Ge *LMM* Auger spectrum of Figure 3.3 show components attributable to Ge^0 and Ge^{4+} separated by over 8 eV.

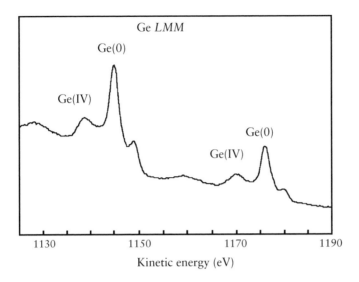

Figure 3.3 Auger chemical state information for a germanium single crystal with a thin layer of oxide

3.2.3 The Auger parameter

The cornerstone of any spectral analysis which relies on peak position to provide information presupposes the ability to determine such values with the necessary accuracy, at least ± 0.1 eV in electron spectroscopy. The two possible sources of error are those due to spectrometer calibration and those resulting from electrostatic charging of the specimen. The former is easily overcome by accurate calibration of the spectrometer against known (standard) values for copper and gold. The latter is resolved for metallic specimens by proper mounting procedures but for insulators and semiconductors a slight shift as a result of charging is always a possibility. Although it is possible to use an internal standard, such as the adventitious carbon 1s position, this is not particularly accurate and will vary slightly with the form and amount of carbon. A much more attractive method is to make use of the chemical shift on both the Auger and photoelectron peak in an XPS spectrum and to record the separation of the two lines; this quantity is known as the Auger parameter (α) and is numerically defined as the sum of the peaks.

$$\alpha = E_B + E_K$$

where E_B is the binding energy of the most intense photoelectron emission peak and E_K is the kinetic energy of the Auger transition. The measured value will thus be independent of any electrostatic charging of the specimen. The elements that yield useful Auger parameters in conventional (AlKα) XPS include F, Na, Cu, Zn, As, Ag, Cd, In, and Te, when using high-energy XPS the list can be extended to include Al, Si, P, S, and Cl through to Ti and V. An example of the Auger parameter using conventional XPS is shown in Figure 3.4, which illustrates the spectral features contributing to the F 1s–F *KLL* Auger parameter. As well as providing chemical state information the Auger parameter is, in some cases, able to provide information on crystal structure and relaxation energies. The quantity is more strictly known as the final-state Auger parameter and is dominated by the final-state relaxation energy. A useful relationship is that the change in Auger parameter relative to the standard state is twice the relaxation energy.

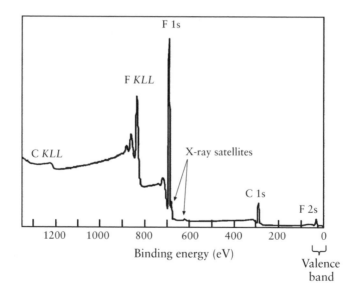

Figure 3.4   Survey XPS spectrum of PTFE showing *KLL* Auger peaks of fluorine and carbon

It is also possible to define an initial-state Auger parameter (ζ) which requires a knowledge of the kinetic energy of the Auger peak and the binding energies of both the levels involved in the de-excitation process. Thus for an *ijj* Auger peak the ζ Auger parameter is defined as:

$$\zeta = E_{K(ijj)} + E_{B(i)} + 2E_{B(j)}$$

This value is less secure than the α parameter described above as it presupposes accurate charge referencing as there is only one kinetic energy term but three binding energy terms, and thus the contribution from any electrostatic charging or charge referencing problems do not cancel out exactly. In the case of the final-state Auger parameter (α) any electrostatic charging will lead to an increase in the binding energy of the photoelectron peak but an apparent decrease in the kinetic energy of the Auger electrons. These contributions are of the same magnitude and thus have no effect on the value of the Auger parameter. In the case of the initial-state Auger parameter (ζ) the magnitude will be distorted by the additional binding energy terms.

3.2.4 Chemical state plots

It is often convenient to use a graphical representation of the final state Auger parameter (α). As stated above, any electrostatic charging present in the spectrum of insulators will readily be accommodated by the use of the Auger parameter. By plotting the kinetic energies of the Auger electrons and the binding energies of the photoelectrons on orthogonal axes it is possible to construct a diagonal grid representing the Auger parameters, equivalent data points along such diagonal lines represent equal values of α. Such a plot is often referred to as a Wagner Plot in recognition of Dr. C. D. Wagner who first presented data in this manner. A Wagner Plot for arsenic compounds using As 3d and As *LMM* peaks is shown in Figure 3.5, and it can be seen that the range of As 3d binding energies is smaller (7 eV) than the complementary range of Auger kinetic energies (11 eV), reflecting the larger magnitude of the Auger chemical shift for this element.

In cases where an Auger peak is not readily apparent in the spectrum, the bremsstrahlung radiation from an achromatic source may be of use.

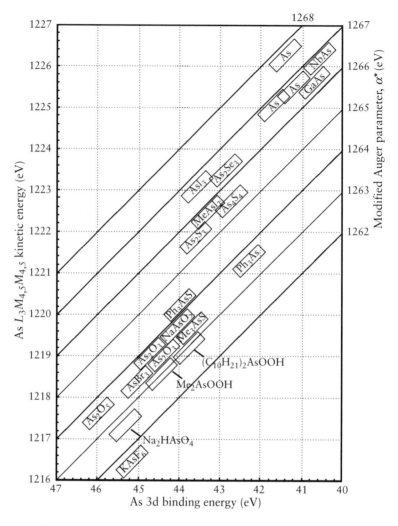

Figure 3.5   Wagner chemical state plot for arsenic compounds

In this manner, it is possible to excite Al *KLL*, Si *KLL*, S *KLL* and Cl *KLL* Auger peaks, for example. In practice this is readily achieved by setting the kinetic-energy scale of the spectrometer to the region of interest, the low intensity of these bremsstrahlung-induced features is not a serious drawback, as the electron background in this region of the spectrum (effectively the negative binding energy region of the XPS spectrum) is very low.

Within the high-resolution spectra of individual core levels there may exist fine structure that gives the electron spectroscopist additional information concerning the chemical environment of an atom. The major features in this category are 'shake-up' satellites and multiplet splitting.

3.2.5 Shake-up satellites

Shake-up satellites may occur when the outgoing photoelectron simultaneously interacts with a valence electron and excites it (shakes it up) to a higher-energy level; the energy of the core electron is then reduced slightly giving a satellite structure a few electron volts below (but above on a binding-energy scale) the core level position. Such features are not very common, the most notable examples being the 2p spectra of the   d-band metals and the bonding to anti-bonding transition of the π molecular orbital ($\pi \rightarrow \pi^*$ transition) brought about by C 1s electrons in aromatic organics. The former is best illustrated by the Cu 2p spectrum; a strong shake-up satellite is observed for CuO, as shown in Figure 3.6, but is absent for Cu_2O and metallic copper. In the case of these features of inorganic compounds it is now appreciated that the dominant factor in the generation of these satellites is final-state effects, such as the screening of the core hole by valence electrons, the relaxation of electron orbitals and the polarization of surrounding species. An allied feature is the 'shake-off' satellite where the valence electron is ejected from the ion completely. These are rarely seen as discrete features of the spectrum but more usually as a broadening of the core level peak or contributions to the inelastic background.

3.2.6 Multiplet splitting

Multiplet splitting of a photoelectron peak may occur in a compound that has unpaired electrons in the valence band, and arises from different spin distributions of the electrons of the band structure. This results in a doublet of the core level peak being considered; multiplet splitting effects

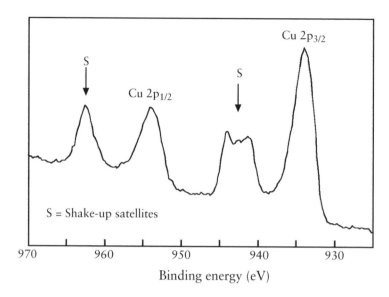

Figure 3.6   Shake-up satellites for Cu 2p spectrum from CuO

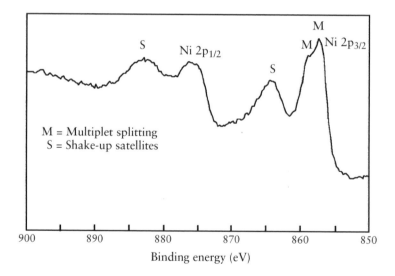

Figure 3.7   Multiplet splitting for the Ni $2p_{3/2}$ spectrum of NiO

are observed for Mn, Cr (3s levels), Co, Ni ($2p_{3/2}$ levels), and the 4s levels of the rare earths. The $2p_{3/2}$ spectrum of nickel shows multiplet splitting for NiO, as shown in Figure 3.7, but not for $Ni(OH)_2$ – a feature that has proved very useful in the examination of passive films on nickel.

3.2.7 Plasmons

The final type of loss feature to be considered is that of plasmon losses. These occur in both Auger and XPS spectra and are specific to clean metal surfaces. They arise when the outgoing electron excites collective oscillations in the conduction band electrons and thus suffers a discrete energy loss (or several losses in multiples of the characteristic plasmon frequency, about 15 eV for aluminium). The characteristic plasmon loss peaks for clean aluminium are shown in Figure 3.8.

The loss features described above can provide valuable information but some, as in the case of plasmons, merely serve to complicate the spectrum. In either case, it is important to assign them correctly so that all spectral features are accounted for and all elements identified before beginning the calculation for a quantitative surface analysis.

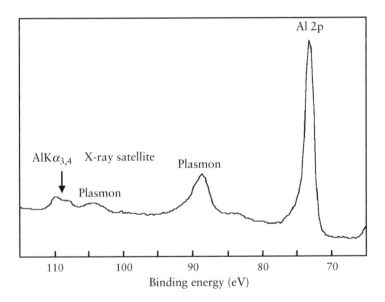

Figure 3.8 Plasmon loss features from clean aluminium

3.3 Quantitative Analysis

In order to quantify spectra from either XPS or AES, one must convert peak intensities (peak areas or peak-to-peak intensities in Auger spectra displayed as differentials) to atomic concentrations. In this section, quantification will be primarily concerned with homogeneous samples. The situation is more complicated for samples which have films at their surface which are either thinner than the information depth of the technique or discontinuous.

3.3.1 Factors affecting the quantification of electron spectra

There are many factors which must be considered when attempting to quantify electron spectra, these are either sample-related factors or spectrometer-related factors.

Sample related factors include the following.

- The cross-section for emission, which is the probability of the emission of an electron due to the effect of the incoming radiation (X-ray photon in XPS or electron in AES). The cross-section depends upon a number of factors such as

  o the element under investigation,

  o the orbital from which the electron is ejected,

  o the energy of the exciting radiation.

In general, for XPS measurements, the cross-section for photoelectron emission from a given orbital quantum number increases with atomic number for a given series of core levels, such as 1s, 2p, etc. For AES the cross-section depends upon atomic number in a more complex manner, passing through a maximum for each transition type (*KLL*, *LMM*, *MNN*, etc.).

- The escape depth of the electron emitted from the atom which depends upon

  o its kinetic energy (the escape depth passes through a minimum with increasing kinetic energy, the minimum occurs in the region of 20 to 50 eV),

  o the nature of the sample.

Spectrometer related factors include the following.

- The transmission function of the spectrometer, which is the proportion of the electrons transmitted through the spectrometer as a function of their kinetic energy. In modern electron spectrometers, this function can be complex and depend upon the way in which the lenses are operated. The transmission function is usually measured by the manufacturers and account is taken of it automatically when data are quantified using the manufacturer's data system.

- The efficiency of the detector, the proportion of the electrons striking the detector which are detected.

- Stray magnetic fields which affect the transmission of low-energy electrons to a greater extent than high-energy electrons and so must be taken into account in the quantification. The effects of magnetic fields are minimized by the use of μ-metal screening around the sample area or by fabricating the analysis chamber in μ-metal.

There are two basic approaches which may be taken in carrying out the calculations for a quantitative evaluation of surface composition: based on first principles, or based on an empirical relationship together with cross-sections or sensitivity factors which may be published or determined in-house. As the quantification of an XPS spectrum is rather more straightforward and potentially more accurate we shall consider it first.

3.3.2 Quantification in XPS

The intensity (I) of a photoelectron peak from a homogeneous solid is given, in a very simplified form, by:

$$I = J\rho\sigma K\lambda$$

where J is the photon flux, ρ is the concentration of the atom or ion in the solid, σ is the cross-section for photoelectron production (which depends on the element and energy being considered), K is a term which covers all of the instrumental factors described above, and λ is the electron attenuation length. The intensity referred to will usually be taken as the integrated area under the peak following the subtraction of a linear or S-shaped background. The above equation can be used for direct quantification (the so-called first principles approach) but, more usually, experimentally determined sensitivity factors (F) will be employed. The parameter F includes the terms σ, K, and λ, in the standard equation, as well as additional features of the photoelectron spectrum such as characteristic loss features. Once a set of peak areas has been calculated for the elements detected, I in the above equation has been determined. The terms σ, K, and λ are incorporated into a set of sensitivity factors appropriate for the spectrometer used, or explicitly incorporated into the algorithm used for quantification. If the X-ray flux remains constant during the experiment (as it invariably does) we can determine the atomic percentage of the elements concerned, by dividing the peak area by the sensitivity factor and expressing it as a fraction of the summation of all normalized intensities:

$$[A] \text{ atomic } \% = \{(I_A/F_A)/\Sigma(I/F)\} \times 100\%$$

The calculation of surface composition by this method assumes that the specimen is homogeneous within the volume sampled by XPS. This is rarely the case, but even so the above method provides a valuable means of comparing similar specimens. For a more rigorous analysis angular dependent XPS may be employed to ensure lateral homogeneity and to elucidate the hierarchy of overlayers present.

3.3.3 Quantification in AES

The quantitative interpretation of Auger spectra is not so straightforward. The first problem encountered is the form of the spectrum. In the differential mode the intensity measurement is the peak-to-peak height. For low-resolution spectrometers, this is approximately proportional to peak area; for high-resolution studies, fine structure, which becomes apparent in the spectrum, reduces apparent peak-to-peak height. It is for this reason that the integrated peak area of a direct energy spectrum is often preferred for quantitative AES. The relative peak areas in a spectrum will depend on the primary beam energy used for the analysis and also the composition of the specimen. It is the latter, matrix, effect that has prevented the production of a series of AES sensitivity factors of the type widely used for XPS. Instead, it is necessary to fabricate binary or ternary alloys and compounds of the type under investigation to provide calibration by means of a similar Auger spectrum; however, the sensitivity factors produced have a narrow range of applicability. In this manner, it is possible to determine the concentration of an element of interest (N_A) as follows:

$$N_A = I_A/(I_A + F_{AB}I_B + F_{AC}I_C + \ldots)$$

where I is the measured intensity of the element represented by the subscript, and F is the sensitivity factor determined from the binary standard such that:

$$F_{AB} = (I_A/N_A)/(I_B/N_B)$$

Various semi-quantitative methods are employed by laboratories throughout the world which relate a measured Auger electron intensity to that of a standard material under the same experimental conditions; this seems to be a fairly satisfactory approach where the time and expense of producing the relevant standard specimens is not warranted.

Although a surface analytical study may be an end in itself, knowledge of the concentration of elements near to the surface is often required. To achieve this, some form of compositional depth profiling is required, either by destructive or non-destructive means and this adds another degree of complexity to the interpretation of the resultant spectra, as we shall see later.

4 Compositional Depth Profiling

Although both XPS and AES are essentially methods of *surface* analysis, it is possible to use them to provide compositional information as a function of depth. This can be achieved in three ways:

- by manipulating the Beer–Lambert equation (Chapter 1) either to increase or to decrease the integral depth of analysis non-destructively (by changing the geometry of the experiment, or the energy of the emitted electron and hence the information depth);

- by removing material from the surface of the specimen *in situ* by ion sputtering – analysis is then alternated with material removal and a compositional depth profile gradually built up;

- by removing material mechanically and examining the freshly exposed surface – common methods for doing this are angle lapping and ball cratering.

4.1 Non-destructive Depth Profiling Methods

4.1.1 Angle resolved electron spectroscopy

These methods are used almost exclusively in photoelectron spectroscopy. Although the principles are equally applicable to Auger electron analysis, the results obtained with the high lateral resolution employed in AES and SAM mean that such changes in analysis depth occur in the analysis of parts of the specimen with different orientations to the electron analyser (because of specimen surface roughness). Such effects tend to be regarded as experimental artefacts to be circumvented by the Auger microscopist, and have only recently become the subject of rigorous scientific investigation.

If we consider the Beer–Lambert equation, discussed in Chapter 1, it is clear that the depth of analysis is dependent on the electron angle of emission, θ[5]. By recording spectra with good angular resolution at a high value of θ, say 75° (relative to the sample normal), an analysis is recorded which is extremely surface sensitive. As normal electron emission is approached ($\theta = 0°$) so the analysis depth moves towards the limiting value of $\sim3\lambda$. This value is often referred to as the XPS analysis depth although it is, of course, more correctly described by $3\lambda \cos\theta$. The relative sampling depths at different take-off angles are illustrated schematically in Figure 4.1(a). A thin overlayer will give a characteristic angular distribution predicted by the Beer–Lambert expression, as shown in Figure 4.1(b). An island-like distribution will show no angular dependence, thus it is possible to distinguish between these two types of phase distribution with relative ease.

This manner of depth profiling is invaluable for compositional changes that occur very close to the surface and has been employed for studies of thin passive films on metals and surface segregation in polymers. In conventional angle resolved XPS (ARXPS), the angular acceptance range of the spectrometer is reduced by the user to provide the required angular resolution. Clearly, there must be a compromise between angular resolution and sensitivity (acquisition time). Spectra are then collected at each of a number of take-off angles by tilting the specimen. The experimental method is illustrated in Figure 4.2 while Figure 4.3 presents an ARXPS data set for a sample of GaAs with a thin oxide layer at its surface.

It is clear from the montage of As 3d spectra that the oxide peak is dominant at the surface whereas the peak due to As in the form of GaAs is more dominant at near normal analysis angles. This phenomenon is repeated in the gallium spectra (not shown).

---

[5]According to the International Standard, ISO 18115, the angle of emission is measured with respect to the surface normal while the take-off angle is measured with respect to the plane of the surface (see Glossary, page 185). Throughout this text the angle of emission will be used.

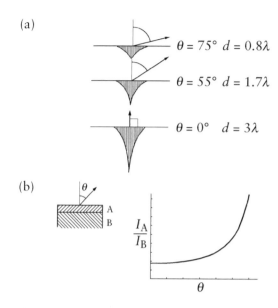

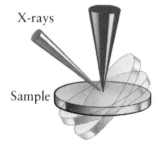

Figure 4.1 Angular electron emission: (a) sampling depth as a function of electron take-off angle (θ) – the width of the shaded areas represent the proportion of the detected electrons emitted as a function of depth; (b) overlayer (A)/substrate (B) and intensity (I) versus θ

Figure 4.2 Conventional ARXPS, using sample tilting

Such information provides a useful guide to the relevant abundance of each element in the near surface layers but there is often a need to provide the thickness of an individual layer. This can be achieved using the Beer–Lambert equation and an ARXPS data set such as that of Figure 4.3, which can be processed to provide such information. It is

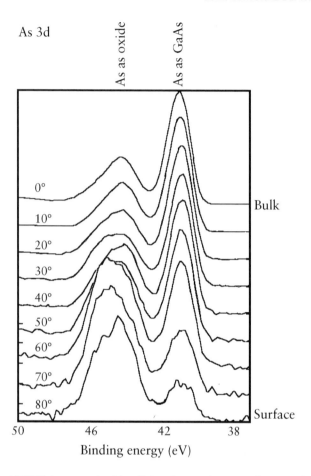

Figure 4.3   ARXPS data acquired by tilting the specimen; in this case the specimen is gallium arsenide

also possible to carry out the converse of the procedure and measure the attenuation length if the overlayer thickness is known.

Measurement of overlayer thickness

Consider a thin layer, thickness d, of material A on a substrate B. To obtain an expression for the signal from A, the Beer–Lambert equation discussed in Chapter 1 must be integrated between 0 and d and becomes:

$$I_A = I_A^\infty [1 - \exp(-d/\lambda_{A,A} \cos \theta)] \qquad (4.1)$$

The signal from B arriving at the B–A interface is I_B^∞, assuming layer B is thick. This signal is then attenuated by passing though layer A. The signal emerging is therefore given by:

$$I_B = I_B^\infty \exp(-d/\lambda_{B,A} \cos \theta) \tag{4.2}$$

Note that the term $\lambda_{B,A}$ is the attenuation length in layer A for electrons emitted from layer B.

Taking the ratio of these signals:

$$\frac{I_A}{I_B} = R = R^\infty \frac{[1 - \exp(-d/\lambda_{A,A} \cos \theta)]}{\exp(-d/\lambda_{B,A} \cos \theta)}$$

where $R^\infty = I_A^\infty / I_B^\infty$

$$R = R^\infty \left[\exp \left(\frac{d}{\lambda_{B,A} \cos \theta} \right) - \exp \left(\frac{d}{\cos \theta} \left[\frac{1}{\lambda_{B,A}} - \frac{1}{\lambda_{A,A}} \right] \right) \right] \tag{4.3}$$

If $\lambda_{A,A} = \lambda_{B,A} = \lambda_A$, which will be approximately true if measurements are being taken from a thin layer of oxide on its own metal, using the same transition, then

$$R = R^\infty \left[\exp \left(\frac{d}{\lambda_A \cos \theta} \right) - 1 \right]$$

Rearranging and taking the natural logarithm

$$\ln[1 + R/R^\infty] = d/(\lambda_A \cos \theta)$$

If it is desired to calculate an equivalent thickness retrospectively, of an oxide for example, it is possible to apply this equation to data recorded at just one value of θ, although this is only appropriate if the acceptance angle of the spectrometer is small. The form of the equation is then:

$$d = \lambda_A \cos \theta \ln[1 + R/R^\infty] \tag{4.4}$$

This approach presupposes that the species attenuating the substrate is a discrete overlayer, which will not always be the case. A more rigorous approach is to record a complete angle resolved data set to

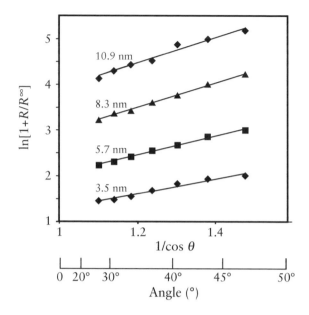

Figure 4.4 Graph showing linearity of the calibration plots for a set of SiO_2 on Si samples which have a range of different oxide thicknesses

test for the presence of a discrete overlayer. This is done by plotting the left-hand side of this equation against $1/\cos\theta$ which will then produce a straight line whose gradient is equal to d/λ_A (as shown in Figure 4.4).

The simple form of this equation can only be used if $\lambda_A \approx \lambda_B$. This is true if the electrons detected from layers A and B have approximately the same energy (e.g., both are emitted from Si 2p). If this is not the case, then the more rigorous Equation (4.3) must be used. The value for R^∞ is the ratio of the intensities of the appropriate peaks from thick samples of the materials. In this context, 'thick' means greater than about 100 nm. The values for the individual intensities will depend upon X-ray flux density, sensitivity factors, atom densities, etc. For a thin layer of SiO_2 on Si most of these factors cancel (assuming the Si 2p peaks are used for both materials) except for the atom densities and the attenuation lengths in the two materials. This means that

$$R^\infty = \frac{\sigma_{Si,SiO_2}\lambda_{Si,SiO_2}}{\sigma_{Si,Si}\lambda_{Si,Si}}$$

where $\sigma_{x,y}$ is the atom number density (atoms per unit volume) of the element x in the material y. Note that R^∞ contains the term $\lambda_{Si,Si}$. The ratio of the atom number densities is given by:

$$\frac{\sigma_{Si,SiO_2}}{\sigma_{Si,Si}} = \frac{D_{SiO_2} F_{Si}}{D_{Si} F_{SiO_2}}$$

where D_x is the density (mass per unit volume) of material x and F_x is the formula weight of x. When there is more than one atom of an element represented by the formula then the formula weight should be multiplied by the number of atoms present. For example, if the number density of oxygen atoms in silicon dioxide is required then F_{SiO_2} should be multiplied by 2.

The following data were collected using ARXPS from a thin layer of silicon dioxide on silicon. Data were collected from a number of samples where the oxide thickness was known, from ellipsometry, and the appropriate graph plotted, using only the angles close to normal emission. The graphs are shown in Figure 4.4. If the gradients of these lines (d/λ) is plotted against the known thickness of the oxide then we can calculate the attenuation length, λ. This is shown in Figure 4.5. From

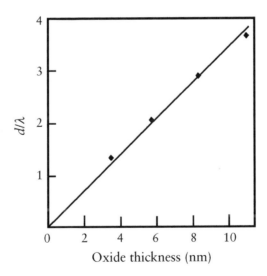

Figure 4.5 The gradients in Figure 4.4 are shown plotted against the known oxide thickness to allow the calculation of the attenuation length, λ

these data λ is calculated to be 3.4 nm. This value can now be fed back into the expression for an unknown sample to calculate the thickness of the oxide.

This methodology can be readily extended to accommodate multilayer samples by the simple expedient of defining a Beer–Lambert term for each layer with (where appropriate) an attenuation term to allow for the reduction of intensity, compared with an infinitely thick standard, of its distance below the surface (Equation (4.2)) and its reduced thickness (Equation (4.1)).

There are, of course, limits in terms of the thickness of an overlayer that can be resolved using ARXPS. If the average thickness of the overlayer becomes comparable with atomic dimensions then the analysis given here is inappropriate because the overlayer would be incomplete. The lower limit for which the thickness measurement can be used is therefore about 0.2 nm. This situation can be accommodated using a different analysis method. At the other end of the scale, the signal generated by the substrate becomes weak even when the electrons are collected at near normal emission angles. In general, the thickest layer which can be analysed using this method is about 3λ which, for silicon dioxide, is approximately 10 nm, (see Figure 4.5).

4.1.1.1 Elastic scattering

If the data in Figure 4.4 are plotted over a wide angular range then the linearity breaks down (see Figure 4.6). This is a result of the effects of elastic scattering of the electrons which are emitted from deeper regions of the sample.

The mechanism is that electrons originating from deep within the sample suffer an elastic collision within the sample and are emitted from the surface at some large angle with respect to the surface normal (Figure 4.7). Because the scattering is elastic, the electrons contribute to the peak intensity. In the case of a thin oxide layer on a metal, the effect causes the signal from the metal to be greater than expected and can lead to layer thicknesses being underestimated. However, good linearity on graphs such as those in Figure 4.6 will generally be expected if the maximum emission angle is restricted to $\sim 60°$.

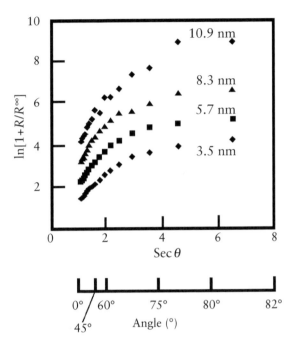

Figure 4.6 The lack of linearity at high emission angles is due to elastic scattering

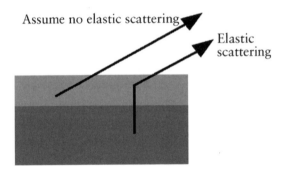

Figure 4.7 Illustration of elastic scattering

4.1.1.2 Compositional depth profiles by ARXPS

Polymers present some special problems in surface analysis in that they are not generally amenable to analysis by AES or ion beam compositional depth profiling because of sample charging and degradation problems. Consequently, angle resolved XPS is one of the few ways of probing near-surface compositional gradients. There are many approaches which can be taken to model such profiles but work from the National Physical Laboratory in the UK has produced validated software in the form of a spreadsheet which incorporates several ways of handling data, the choice depends on the type of information which is desired from the ARXPS data set. Named ARCtick, the routine is fully described at http://www.npl.co.uk/npl/cmmt/sis/arctick.html, and an example of the output from ARCtick is provided in Chapter 5.

Computational methods

No unique transformation from angle-dependent intensities to depth-dependent concentration exists. This implies that a least squares fit of trial profiles to experimental data is not sufficient to determine the accurate concentration profiles. The concept of maximum entropy has therefore been introduced to produce a smooth profile, avoiding the 'over fitting' which a method based on least squares fitting would produce.

In outline the procedure is as follows. A random depth profile is generated and the ARXPS intensities expected from such a profile are calculated. The profile consists of a number of finite layers in which the concentrations of the various elements are calculated. It is convenient to restrict the number of layers to ~10–20, this value is a compromise between depth resolution and computational time. The calculated ARXPS data are compared with the experimental data and the error is calculated:

$$\chi^2 = \sum \frac{(I_k^{calc} - I_k^{obs})^2}{\sigma_k^2}$$

where σ is the standard deviation. The entropy term (S) is then calculated from the trial profile:

$$S = \sum_j \sum_i c_{j,i} - c_{j,i}^0 - c_{j,i} \log \left(\frac{c_{j,i}}{c_{j,i}^0} \right)$$

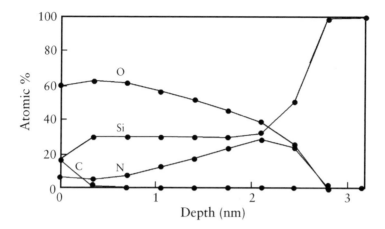

Figure 4.8 Depth profile through an ultra-thin layer of silicon oxynitride reconstructed from ARXPS data using the maximum entropy method

The quantity $c_{j,i}$ is the concentration of element i in layer j. The maximum entropy solution is derived by minimizing χ^2 while maximizing the entropy. This can be done by maximizing the joint probability function, Q:

$$Q = \alpha S - 0.5\chi^2$$

where α is a regularizing constant, providing a suitable balance between the least squares term and the entropy term. Concentration space is then searched to find the optimum value of Q, to reduce the computational time.

This method has been successful in the reconstruction of depth profiles from layers whose thickness is less than about twice the attenuation length of electrons within the layer. An example of the use of this method is shown in Figure 4.8 for a layer of silicon oxynitride on silicon. In this example, the distribution of oxygen and nitrogen within the ultra-thin layer is clearly seen.

4.1.1.3 Recent advances in ARXPS

There have been two recent advances which ensure that ARXPS will become more widely used in future years. The first is a greater experimental appreciation and theoretical understanding of elastic scattering, which has enabled the phenomenon to be treated rigorously in ARXPS algorithms. This has enabled generalized procedures, applicable to many classes of specimen, to be documented, and these are now being promulgated as the ARCtick ARXPS software as indicated in Section 4.1.1.2 above.

The other important advance concerns the hardware available to record ARXPS data. Conventional ARXPS is achieved by the simple expedient of tilting the specimen so that the angle subtended by the sample normal and electron analyser optics is increased in increments of 10–15°. This approach to the technique is illustrated in Figure 4.2. ARXPS applied to large samples is not so straightforward because it would be difficult to tilt a large sample in a typical XPS system, especially if data a required from a region near the edge of the sample. An alternative approach, recently introduced, is the use of the focusing properties of the transfer lens and the hemispherical analyser in conjunction with a position-sensitive detector to acquire parallel (rather than serial) ARXPS profiles. In this manner the angle-resolved data set is acquired simultaneously over a 60° range without the need for mechanical movement of the sample, as described in Chapter 2.

This route to ARXPS has a number of significant advantages compared with the conventional approach.

1. ARXPS can be applied readily to large samples without the difficulties mentioned above.

2. Small area ARXPS is possible but would be very difficult using conventional methods for the following reasons. First, at all angles, the analysis position would have to be accurately aligned with the feature to be analysed. This is difficult, especially if the analysis position is at some distance from the tilt axis and the required analysis area is very small. Second, the analysis area changes as a function of angle, as can be seen in Figure 4.9. A worst case occurs when the transfer lens is used to define the analysis area. Using parallel angle acquisition, the analysis area and position is completely independent of the emission angle.

3. If an insulating sample is tilted the charge compensation conditions also change. Changes in peak position or shape may then be due to changes in the efficiency of charge compensation. Using parallel

Figure 4.9 The analysis area changes as a function of angle, especially when using lens-defined small area analysis

acquisition of angular data, the compensation conditions are the same for all angles and any changes in the spectra as a function of angle must reflect real chemical differences.

4.1.2 Variation of analysis depth with electron kinetic energy

An alternative way of obtaining in-depth information in a non-destructive manner is by examining electrons from different energy levels of the same atom. The inelastic mean free path varies with kinetic energy, and by selecting a pair of electron transitions which are both accessible in XPS but have widely separated energies, it is possible to obtain a degree of depth selectivity. The Ge 3d spectrum (kinetic energy = 1450 eV, $\lambda \sim 2.8$ nm) of Figure 4.10(a) shows Ge^0 and Ge^{4+} components with the oxide component being about 80 per cent of the elemental. If we also record the spectrum of the Ge $2p_{3/2}$ region (kinetic energy = 260 eV, $\lambda \sim 0.8$ nm) we see the elemental component is merely a shoulder on the Ge^{4+} peak (Figure 4.10(b)), thus confirming the presence of the oxide layer as a surface phase.

It is possible to obtain a similar effect by using the same electron energy level but exciting the photoelectrons with a series of different

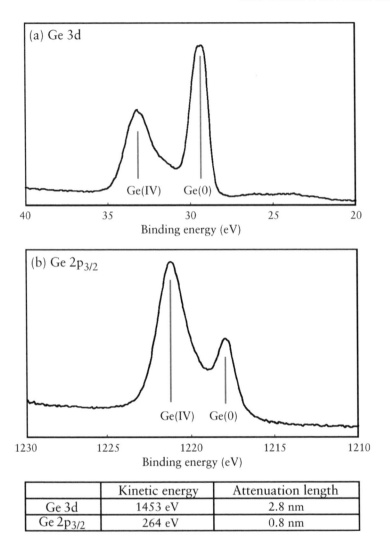

Figure 4.10   Ge 3d and Ge $2p_{3/2}$ spectra showing variation of sampling depth with electron kinetic energy

X-ray energies. For instance, C 1s electrons have a constant binding energy of 285 eV, which corresponds to a kinetic energy of 969 eV in MgKα radiation, 1202 eV in AlKα, and 2700 eV in AgLα radiation. These kinetic energies yield analysis depths of approximately 6, 7, and 10 nm for a typical polymer. Although high-energy X-ray sources for

XPS are still rare, the conventional Al/Mg twin anode fitted to most spectrometers does provide a modest depth profiling capability which is often sufficient to distinguish between a surface layer and an island-like distribution.

4.2 Depth Profiling by Erosion with Noble Gas Ions

4.2.1 The sputtering process

Although the non-destructive methods described above are extremely useful for assessing compositional changes in the outer 1–10 nm of material, to obtain data from depths greater than this it is necessary to remove material by ion bombardment within the spectrometer.

The literature available on the subject of ion beam–solid interactions is enormous (see Bibliography for examples). All that can be achieved here is to make the reader aware of general principles and the possible causes of profile distortion. The primary process is that of sputtering surface atoms to expose underlying atomic layers. At the same time, some of the primary ions are implanted into the substrate and will appear in subsequent spectra. Atomic (cascade) mixing results from the interaction of the primary ion beam with the specimen and leads to a degradation of depth resolution. Enhanced diffusion and segregation may also occur and will have the same effect. The sputtering process itself is not straightforward; there may be preferential sputtering of a particular type of ion or atom. Ion-induced reactions may occur; for instance copper (II) is reduced to copper (I) after exposure to a low-energy low-dose ion beam. As more and more material is removed so the base of the etch crater increases in roughness and eventually interface definition may become very poor indeed.

A high-quality vacuum is essential if a good depth profile is to be measured. If there are high partial pressures of reactive impurities present, the surface which is analysed may not reflect the material composition. This arises because the act of sputtering produces a highly reactive surface which can getter residual gases from the vacuum. Oxygen, water and carbonaceous materials are common contaminants. For this reason, also, the gas feed to the ion gun must be free from impurities.

4.2.2 Experimental method

Figure 4.11 shows a flow chart which illustrates the experimental procedure used to obtain a depth profile. The experiment usually begins with an analysis of the undisturbed surface using either XPS or AES. The sample then undergoes a period of sputtering (ion etching) using ions whose energy is in the range of a few hundred to a few thousand eV. Following this, the ion beam is then switched off (blanked) and the sample is analysed again. This process is continued until the required depth is reached. When an insulating sample is being profiled, the steady-state surface potential during the etching period will be different from that during the analysis. This could cause peak shifts during the early part of the analysis. To overcome this, a settling time can be used during which period the sample is exposed to the analysis conditions but no data are collected. This procedure allows the surface potential to regain its steady-state condition before attempting to collect chemical state information.

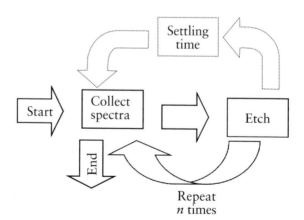

Figure 4.11 Flow diagram showing the experimental method for producing a depth profile in either AES or XPS; the settling time is optional and can be used for deriving profiles from insulating samples

The current density profile of an ion beam is generally not uniform – many approximate to a Gaussian cross-section. Such a beam cross-section would produce a crater in the sample which does not have a flat bottom. Poor depth resolution would result from extracting data from such a crater. To overcome this difficulty, the ion beam is usually scanned or rastered over an area which is large with respect to the diameter of the beam. Rastering produces a crater which has a flat area at the centre from which compositional data can be obtained.

When collecting the spectroscopic data from the crater it is important to limit the data acquisition to the appropriate area within the crater, avoiding the area close to the walls where the crater bottom is not flat. This is usually a simple matter in AES because the electron beam used for the analysis usually has a much smaller diameter than the ion beam used for etching the sample. The electron beam can therefore be operated in point analysis mode in the centre of the crater or rastered over a small area. For XPS, the methods of small area analysis are generally used, as described in Chapter 2. Figure 4.12 shows an example of a simple XPS depth profile through a tantalum oxide layer grown on tantalum metal.

There are many points to bear in mind when selecting the experimental conditions for a depth profile. Most of these are concerned with

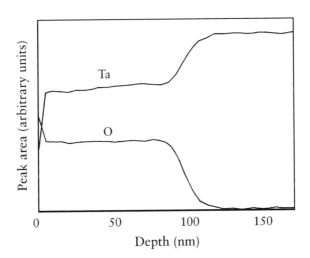

Figure 4.12 XPS depth profile through a layer of tantalum oxide on tantalum metal

either the speed of acquisition or the depth resolution; the requirements for speed are usually opposite to those for resolution. It is important to understand the principles which govern profiling rate and resolution.

4.2.3 Sputter yield and etch rate

The sputter yield is the physical property which determines the rate at which material is removed from the sample during bombardment by energetic ions. The definition of sputter yield is:

Sputter yield (Y) = Number of atoms removed/Number of incident ions

This is the physical property of a sample which has the greatest influence upon the rate at which it can be profiled. To be of practical value the sputter yield needs to be converted to an etch rate usually given in the units of $nm\,s^{-1}$.

Rate of removal of surface atoms = $Y \times$ Rate of arrival of ions

The rate of arrival of ions is given by:

$$I/e$$

where I is the ion beam current in amps and e is the charge on the ion in coulombs, we now have:

Rate of removal of surface atoms = $(YI)/e$

To convert this to a depth scale, we need to know the number of atoms under the beam's rastered area and the atomic layer thickness, which can be calculated from the atomic weight and the density of the sample:

Number of atoms cm^{-3}

$$= (\text{Target density} \times \text{Avogadro's number})/\text{Atomic weight}$$
$$= (\rho N)/w$$

where ρ = density, N = Avogadro's number, w = atomic weight.

If we make the approximation that the atoms are in a cubic array then the number of atoms in 1 cm^2 is $(dN/w)^{2/3}$. Thus, if the beam is rastered over an area of A cm^2, the number of surface atoms in the rastered area is $A([dN]/w)^{2/3}$. Similarly, the layer thickness is $[w/(dN)]^{1/3}$. Now, the number of atomic layers removed per second is the number of atoms removed per second divided by the number of atoms at the surface under the ion beam:

$$\text{Layers per second} = \frac{IY}{e}/[A(\rho N/w)^{2/3}] \tag{4.5}$$

To get to the etch rate we have to multiply this expression by the layer thickness:

$$\text{Etch rate} = \left(\frac{w}{\rho N}\right)^{1/3}\left[\frac{IY}{e}\right]/[A(\rho N/w)^{2/3}]\,\text{cm s}^{-1} \tag{4.6}$$

This simplifies to:

$$\text{Etch rate} = IYw/(Ae\rho N)\,\text{cm s}^{-1} \tag{4.7}$$

For greater convenience, if the etch rate is expressed in nm s^{-1}, the beam current in µA, area in mm and eN as 10^5 coulombs (a good approximation) then the expression becomes:

$$\text{Etch rate} = 10^{-2}IYw/(\rho A)\,\text{nm s}^{-1} \tag{4.8}$$

4.2.4 Factors affecting the etch rate

Material

The sputter rate depends upon the chemical nature of the material, not only the elements present but also their chemical state. It is difficult to predict the sputter yield for a material but there are a number of computer simulations available. Some of these can predict sputter yields of elements with reasonable accuracy but they become less reliable when applied to compounds or alloys. It is usually preferable to measure the sputter yield experimentally under the conditions normally employed.

Ion current

From Equation (4.8), above, it can be seen that the etch rate is directly proportional to the ion current. It should therefore be possible to increase the etch rate by using the maximum beam current available. However, in a normal ion gun the spot size of the beam increases with increasing beam current and so the rastered area must be increased to ensure that the crater bottom remains flat. With this in mind, simply increasing the beam current will not necessarily increase the etch rate. It is the ion beam current density which is the important parameter rather than merely the current.

Ion energy

In the energy range normally employed in XPS and Auger profiling the sputter yield increases with ion energy. At high energy, the sputter yield will reach a maximum. Higher energies also mean smaller spot sizes at a given beam current and so will lead to better crater quality. The higher etch rate will, however, be accompanied by poorer depth resolution because the ions can penetrate deeper into the material causing atomic mixing.

Nature of the ion beam

In XPS and Auger profiling, it is customary to use the ions of the noble gases for sputtering. In the energy range which is normally used, sputter yield increases with increasing atomic mass; xenon ions provide higher etch rates than argon and helium ions provide very much lower sputter yields than argon. In addition, the larger ions penetrate a shorter distance into the material and therefore allow better depth resolutions to be obtained. This all points to the use of xenon as the ion beam but, because of its significantly higher cost, xenon is seldom used; in practice argon is the gas which is selected.

Angle of incidence

As the angle of incidence (measured from the sample normal) is increased, the sputter yield increases reaching a maximum at about 60°. Above this angle it decreases rapidly. The way in which the sputter yield varies with angle is difficult to predict because it depends upon the material being sputtered and the nature of the ion beam. However, since an angle of 60° provides high sputter yields and good depth resolution, many commercial instruments are designed with the ion incidence angle close to this angle.

4.2.5 Factors affecting the depth resolution

Depth resolution in a depth profile is a measure of the broadening of an abrupt interface brought about by physical or instrumental effects. A commonly accepted method for measuring depth resolution is to measure the depth range, Δz, over which the measured concentration changes from 16 per cent to 84 per cent of its total change while profiling through an abrupt concentration change (see Figure 4.13).

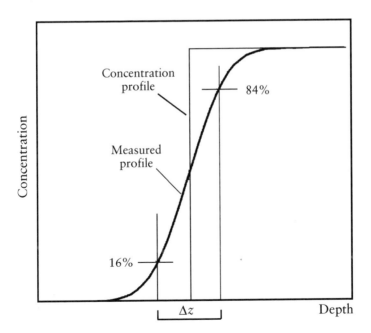

Figure 4.13 Definition of depth resolution

Many of the factors which affect sputter rate also determine the depth resolution available. Some of these factors relate to the characteristics of the sample, some to the instrument and some to the physical process of sputtering.

Ion beam characteristics

The extent to which the ion beam characteristics affect depth resolution generally relates to the depth range of the ions after striking the sample. This is because the passage of an energetic ion through a solid causes atomic mixing along the whole trajectory of the ion. Hence low ion energy, grazing incidence angles and heavy ions lead to the best depth resolution because these minimize the depth range over which mixing can occur.

Crater quality

The crater must be as flat as possible over the analysis area. If this is not the case then information is being collected from a range of depths and resolution suffers. Generally, the raster dimensions should be at least five ion beam diameters to get good flatness over a reasonable distance within the crater.

Beam impurities

Chemical impurities in the beam can be minimized by using a high-purity gas feed. Other impurities are more difficult to remove; these consist of high-energy neutral species and ions with a multiple charge.

High-energy neutrals are formed by the collision of an energetic ion with a gas atom during which there is a charge exchange process. Following this process, the neutral species continue with their kinetic energy almost unaltered but without their positive charges. Neutral species cannot be focused or scanned and so they can sputter the sample in an undefined manner and disrupt the quality of the crater. The concentration of neutral species in the beam can be minimized by providing effective pumping near the source region of the ion gun. This means that the high-energy ions do not have a long path through the high-pressure region, minimizing the probability of collision with a gas atom. Some ion guns place a bend in the ion trajectory. The ions can be deflected electrostatically to follow the bend while the neutral species cannot and so they do not reach the sample.

An ion having a double positive charge will have twice the kinetic energy of an ion with only a single charge. The higher energy will cause greater penetration of the sample and therefore adversely affect the depth resolution. The fraction of ions with multiple charges will depend upon the conditions in the source of ions.

Information depth

As we have seen earlier, electrons are collected from a range of depths, not just the top monolayer of material. The depth from which the electrons are collected will affect the depth resolution. In electron spectroscopic techniques the lower the kinetic energy of the collected electrons the smaller is the information depth and therefore the better the depth resolution. This phenomenon is illustrated in Figure 4.10 which shows spectra recorded from a germanium sample which has a thin oxide layer at the surface. The spectrum from the 3d region shows a large metallic peak whereas the spectrum from the 2p region has a much smaller contribution from the metallic peak. This is because the kinetic energy of the electrons forming the 2p spectrum is much lower and therefore a lower proportion of them are able to pass through the oxide layer.

Tilting the sample so that the detected electrons are those which leave the sample surface at more grazing angles can also control information depth. Figure 4.14 shows a comparison of spectra taken at two different angles. The sample was a thin oxide layer on silicon. It is clear that the relative contribution of the oxide to the spectrum is much larger in the spectrum taken at grazing emission angles (spectrum (b)); this is due to the reduced information depth when electrons emitted at grazing emission angles are detected. The disadvantage of collecting at grazing emission is that the signal intensity is lower; the greater noise in the spectrum (b) is evidence for this. The advantage of using grazing emission angles in sputter profiling is that the reduction in the information depth improves the depth resolution.

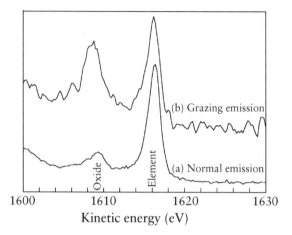

Figure 4.14   Si $KL_{2,3}L_{2,3}$ Auger spectra taken from silicon with a thin oxide layer at the surface; one spectrum (a) was collected along the surface normal and the other (b) was collected at grazing emission (the relative intensity of the oxide peak is clearly much larger in this spectrum)

Original surface roughness

If the surface of the sample is rough then this will affect the overall depth resolution, the roughness being maintained (or worsened) through the profile; information will therefore be collected from a range of depths at any one time.

An initially rough sample surface may become even rougher as sputtering progresses for a number of reasons.

- A rough surface may shadow parts of the sample from the ion beam.

- Rough surfaces may also present many differing crystal facets to the ion beam. Each crystal facet will have a different sputter yield.

- There will be a range of incidence angles between the ion beam and the sample surface. This will produce a range of sputter yields.

Induced roughness

The sputtering process can cause the surface to become rough during the profiling experiment, degrading the resolution as a function of depth. This problem can be significantly reduced or eliminated by rotating the sample beneath the ion beam (azimuthal rotation) during the sputtering cycles. Many commercial instruments now offer sample stages which incorporate azimuthal rotation.

Preferential sputtering

In a multi-component sample the sputter yields from different elements can be different. Under these conditions there will be roughening which may not be controllable by azimuthal rotation. Furthermore, the surface concentration of the elements will be different from the bulk concentration and so quantification of the data will be difficult, requiring the application of correction factors.

Redeposition of sputtered material

Care must be taken to avoid the redeposition of sputtered species onto samples awaiting analysis. Furthermore, if the etch crater is small, material can be sputtered from the crater walls and redeposited within the analysis area.

4.2.6 Calibration

Depth profile data are presented as elemental intensity versus etch time, and a major problem in sputter depth profiling is converting this etch time scale to a depth scale. Although, in special cases, it is possible to calibrate ion guns for a particular material, it is a time-consuming procedure and, more often, the sputter rate is related to an international standard. The current standard is a Ta_2O_5/Ta foil with an accurately determined oxide thickness (30 nm or 100 nm). Thus, it is possible to report a sputter rate and also an interface width for the ion gun and the conditions used in any particular piece of work.

4.2.7 Ion gun design

In addition to providing a means of compositional depth profiling in surface analysis, ion guns may be fitted to a spectrometer for additional purposes such as large area specimen cleaning or as the primary beam in ion beam analysis methods such as ISS or SIMS. The requirements for each application are slightly different and there are several different designs in common use.

The three most widely employed types of ion source for surface analysis are, in order of increasing performance and cost, the cold cathode static spot gun, the electron impact ion source, and the duoplasmatron type of ion gun. Liquid metal ion guns also find use in surface analysis, principally for small area depth profiling and imaging SIMS; their use is rare on XPS and AES instruments.

Cold cathode ion guns

In the cold cathode type of gun a variable potential of $1-10\,kV$ is utilized in conjunction with an external magnet to produce a discharge in the ionization region of the gun into which argon or another inert gas flows to a pressure of about $10^{-6}\,mbar$. The positively charged ions are extracted and the beam shaped by a simple electrostatic lens. These guns generally give a static spot size of $5-10\,mm$ depending on the applied focus potential. They are recommended for specimen cleaning, but with the addition of an aperture assembly below the focus electrode, they can give remarkably uniform etch craters. The main disadvantage is the production of neutral species which are not deflected by the focus potential and give rise to a 'sub-crater' of about $5-10$ per cent of the area of the main crater. The maximum current available with this type of gun is of the order of $50\,\mu A$.

Electron impact ion guns

Ion guns based on the electron impact source are very popular for depth profiling applications in XPS and AES. This is due to a combination of their compact design and lower cost compared with duoplasmatron designs. In addition, the ion current output of this type of ion gun at very low

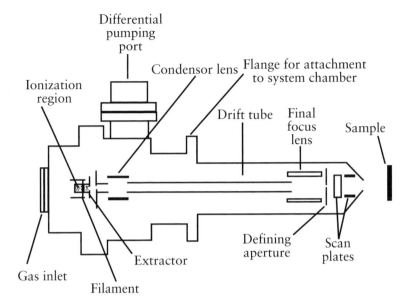

Figure 4.15 Schematic diagram of a typical ion gun with electron impact source

acceleration voltages is usually higher than that from a duoplasmatron, this is important when extremely good depth resolution is required.

In this type of source, electrons from a heated filament are accelerated into a cylindrical grid where they collide with gas atoms, giving rise to the formation of ions. The ions are then extracted from the ionization region. The kinetic energy of the ions is controlled by the magnitude of the potential applied to the grid (up to 5 kV). This ion source produces a small energy spread with only a small fraction of neutrals (i.e., unionized atoms) present in the beam. Spot sizes commercially available vary from 2 mm (at 5 µA) down to 50 µm (at 500 nA). The beam can usually be scanned (rastered) over the surface to produce high-quality craters. A schematic diagram of a typical ion gun of this type is shown in Figure 4.15.

Duoplasmatron ion guns

For the rapid removal of material and etching of large areas, the duo-plasmatron design of ion source is sometimes preferred. A magnetically constricted arc is used to produce a dense plasma from which the ion beam is extracted, focused, and rastered across the specimen by a set of deflector plates. Ion guns based on a duoplasmatron source provide an intense source with a narrow energy spread, making them suitable for small spot focusing. The current density achieved with such an ion gun can be high, leading to high etch rates. Duoplasmatron ion sources are available in a range of spot sizes varying from 2 mm (providing 80 μA of ion current and with a field of view of approximately 15 mm × 15 mm) to better than 5 μm (providing a maximum current of 5 μA and a 2 mm × 2 mm field of view). The former is ideal for large area depth profiling in non-monochromated XPS while the latter finds extensive use as a primary source in the ion beam analysis of materials.

With both duoplasmatron and electron impact sources the beams may be 'purified' (i.e., removal of impurities and ions with a multiple charge) by the addition of a Wien filter (a crossed magnetic and electrostatic field mass separator) and a small deflection within the gun design to eliminate neutrals. Such high-purity beams are not, however, a prerequisite for good quality XPS and AES sputter depth profiling. In general, the only precaution necessary is the provision of a high-purity gas feed.

Liquid metal ion guns

For some applications, particularly when very small diameter ion beams are required, liquid metal ion guns can be used. The metal ions produced by this type of gun are usually Ga^+ but other materials have been used. These guns can produce spot sizes below 50 nm at energies above 25 keV. Although the beam current at these small spot sizes is very small (typically ~50 pA), the current densities are very high and large etch rates can be achieved over a small area. They have the added advantage that they do not impose a gas load on the vacuum system during operation. This type of ion gun is commonly encountered in SIMS analysis, occasionally used for AES profiling and rarely, if ever, used in XPS profiling. A major application of this type of gun is in micromachining, often referred to as focused ion beam (FIB) technology. Examples of its use are:

• for producing a crater in, for example, a processed silicon wafer prior to the analysis of the side wall using AES,

• for repairing lithographic masks used in the production of semi-conductor devices.

4.3 Mechanical Sectioning

To analyse to greater depths than is practical using sputtering, it is necessary to resort to an *ex situ*, mechanical process for removing material. Two related methods will be outlined here, angle lapping and ball cratering.

4.3.1 Angle lapping

When this method is employed, material is removed by polishing the specimen at a very shallow angle ($<3°$) and then introducing the sample, with any buried interface now exposed, into the spectrometer. A brief ion etch to remove contamination is all that is needed prior to analysis. By carrying out Auger point analyses in a stepwise manner the variation of concentration with depth is established and it is a matter of simple geometry to convert the position of the analysis in the x–y plane to the distance from the original surface, the z plane (see Figure 4.16). The main difficulty with this technique is the need to produce a very shallow taper section with a flat surface and well-defined geometry.

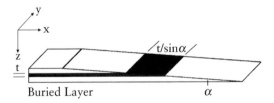

Figure 4.16 A taper section showing how the thickness of a buried layer, t, can be amplified to $t/\sin \alpha$

4.3.2 Ball cratering

The problem of cutting the taper is overcome, to a large extent, by using an allied process known as ball cratering. In this process,

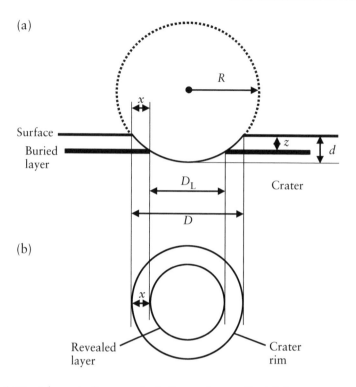

Figure 4.17 Schematic diagram of a ball-cratering profile: (a) shows a cross-section of a crater through a surface and a buried layer, while (b) shows the plan view of the specimen and the revealed sub-surface layer

mechanical sectioning of the specimen is carried out by a rotating steel ball of known diameter (usually about 30 mm) coated with fine (~1 μm) diamond paste which rotates against the specimen and produces a shallow, saucer-like crater. The ball can be removed from time-to-time to assess the progress of the lapping and, on replacement, automatically 'self-centres' in the crater. Figure 4.17 shows a schematic representation of the ball-cratering process.

From knowledge of the radius of the sphere (R) and the crater diameter (D) the crater depth (d) can be calculated as

$$d = \frac{D^2}{4(2R + d)}$$

but, as d is very small compared with R, this approximates to

$$d = D^2/8R$$

If x is the radial distance from the edge of the crater to the revealed buried layer then the depth of the layer, z, can be shown to be:

$$z = \frac{x}{2R}(D - x)$$

By recording Auger point analyses along the surface of the crater a compositional depth profile can be determined. If there are buried interfaces of special interest, ion sputtering may be used at a point on the crater close to the interface to obtain better depth resolution.

Ball-cratering devices, illustrated in Figure 4.18, are available commercially. They consist of a horizontal shaft with a reduced diameter in the form of a 'V'. The sample is mounted near the drive shaft and the hardened-steel ball rests on it, driven by the horizontal shaft. Diamond paste is commonly used as an abrasive. Inspection of the crater, using the integral microscope, is carried out during the erosion process after removing the ball. The ball is readily relocated in the crater if further material removal is required.

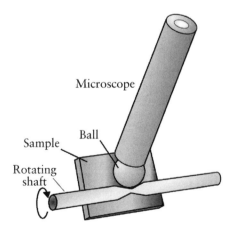

Figure 4.18 Schematic diagram of a ball-cratering device

Ball cratering works well for metals and oxides but there are problems with both soft and certain brittle materials. Polymers are extremely difficult to handle but some success has been obtained by using a ball-cratering machine equipped with a cryostage.

4.4 Conclusions

Sputter depth profiling is by far the most popular means of producing a compositional depth profile in surface analysis. Although the above discussions suggest that it is fraught with difficulties, it is fair to say that the majority of the problems can be circumvented or reduced to an acceptable level by careful experimental techniques. It is for this reason that, as a method for depth profiling, it is widely used in studies of metals, oxides, ceramics, and semiconductors, as we shall see in the next chapter.

The analysis depth which is feasible varies with the sample and the system employed but the upper limit is of the order of a few microns. The main reason for this is the time for the experiment. Sputter profiling becomes unsuitable when the layer thicknesses become either very large or very small. When the thickness is so great that the time for the experiment would be excessive then either angle lapping or ball cratering should be considered.

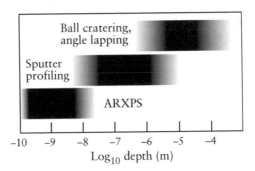

Figure 4.19   The range of depths for which the three types of depth profiling are suitable; the darkest colour indicates greatest suitability

For very thin layers there may be problems associated with the attainment of the so-called sputter equilibrium leading to uncertainties in the sputter yield. Furthermore, the process of sputtering can lead to changes in the chemical composition of the material. Under these circumstances the use of ARXPS has many advantages, bearing in mind that it is only appropriate for layers up to about 10 nm.

Figure 4.19 shows the range of depths for which each of the three types of profiling method is appropriate. The depth ranges will depend, to some extent, upon the material and so Figure 4.19 should be regarded as a guide only.

5 Applications of Electron Spectroscopy in Materials Science

5.1 Introduction

So far, in this text, we have been concerned with the practice of electron spectroscopy and the interpretation of the resultant spectra. We will now consider the way in which it is possible to make use of these surface analysis techniques to provide information which furthers our knowledge in a particular discipline. Although XPS and AES together with SAM are used widely in all branches of pure and applied sciences – as well as for trouble shooting and quality assurance purposes – the only area that we will consider in this chapter is their use in materials science investigations. If we subdivide this group, it is possible to identify the following applications headings: metallurgy (including surface engineering), corrosion, microelectronic materials and devices, polymers, and adhesion. We shall consider each of these areas in turn; representative references for each are listed in the Bibliography which will provide interested readers with further examples and guidance in their particular field.

5.2 Metallurgy

In the field of metallurgy it is Auger electron spectroscopy which has proved to be the most popular technique, and with good reason. The majority of investigations are concerned with the diffusion of elements within metallic matrices. This may take the form of interdiffusion of metallic coatings with the substrate, or the surface segregation of minor alloying elements on heating in oxidizing or reducing atmospheres. However, the major contribution of Auger electron spectroscopy to metallurgy, especially in the early days of the development of surface analysis, was the investigation of grain-boundary segregation and embrittlement in structural steels. In addition, both AES and XPS have been used in 'quality assurance' and sometimes 'forensic' roles to ensure (for example) rolled-steel sheet is of adequate cleanliness, or to identify surface phases which lead to poor compaction in powder metallurgy processing.

5.2.1 Grain-boundary segregation

The embrittlement of structural steels results from the aggregation of certain elements, present in very low or trace quantities in the bulk material, at the prior austenite grain-boundaries. The grain boundaries are weakened to such an extent that they become the preferred fracture path, with catastrophic effects on the material's mechanical integrity. The elements most widely investigated are phosphorus and sulphur but the effect is brought about by, and has been studied for, silicon, germanium, arsenic, selenium, tin, antimony, tellurium, and bismuth. The quantity of grain-boundary segregant involved is necessarily very small, probably sub-monolayer, and located at the grain-boundary within a material of grain size of about 100 μm or less. Thus, the need for surface specificity and reasonable spatial resolution is immediately apparent. In order to measure the quantity of segregant at the interface, the steel must be fractured in an intergranular manner usually at, or near, liquid nitrogen temperature. This must be carried out within the UHV environment of the spectrometer to prevent oxidation of the iron matrix and subsequent obliteration of the small signal from the segregant. Nowadays, most manufacturers offer such a fracture stage for their Auger microscopes, the more sophisticated having the ability to analyse both fracture surfaces (Figure 5.1). All rely on fracture by a fast three point bend configuration (similar to the geometry referred to by the metallurgist as an Izod Test). Scientists requiring controlled strain-rate fracture must still resort to building their own devices.

Figure 5.1   Fracture stage attached to a modern Auger electron spectrometer

The surface morphology generated by such a low-temperature fracture is sometimes a mixture of regions of both intergranular and transgranular failure. This provides a convenient comparison between the matrix composition (transgranular) and a grain-boundary analysis from the region of brittle failure; Auger spectra taken from regions of this type are presented in Figure 5.2. The presence of oxygen indicates that

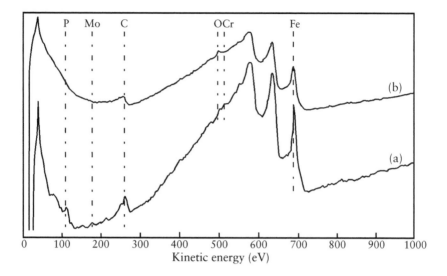

Figure 5.2 Auger spectra from regions of (a) intergranular, and (b) transgranular failure of an alloy steel fractured at liquid nitrogen temperatures

even under clean UHV conditions some contamination of the surface may occur.

Studies have been made of many systems that exhibit grain-boundary segregation and the underlying theory is now well developed, mainly as a result of systematic studies undertaken at the National Physical Laboratory, UK. Thus, the extent of grain-boundary segregation may be predicted by the following equation for dilute levels of segregant in the matrix:

$$\beta = \frac{K}{X_C^0}$$

where β is the grain-boundary enrichment ratio, X_C^0 is the solid solubility of segregant in the matrix and $K = \exp(-\Delta G/RT)$, ΔG is the free energy of segregation. This equation describes a large number of experiments undertaken on many systems all indicating that the degree of enrichment is dependent on solid solubility over a very wide range (from 100 p.p.m. to 100 per cent). These data are presented graphically in Figure 5.3.

Thus in the field of segregation and embrittlement AES has not only provided a technique that enables the level of segregant to be

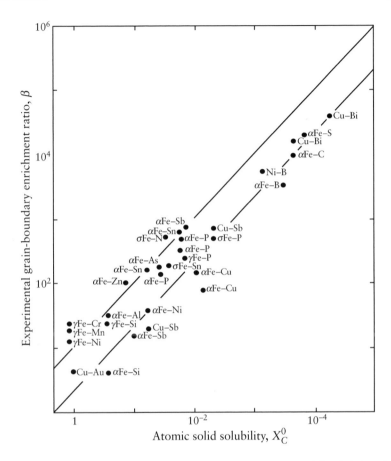

Figure 5.3 Grain-boundary enrichment (β) versus inverse solid solubility for a range of binary alloys (Seah, M. P. and Hondros, E. D. (1997). *Int. Met. Rev.*, **22**, 262–301, Fig. 5.1, reproduced by permission of the Institute of Materials, Minerals and Mining)

qualitatively assessed it has, as a result of such measurements, enabled the development of an underlying theory which predicts such a phenomenon very accurately.

The *in situ* fracture approach can sometimes identify discrete minor phases present in an alloy. The image of Figure 5.4 shows a single crystal precipitate some 3 μm across at the fracture surface of a steel; the Auger spectrum (Figure 5.5) shows this to be a single crystal of aluminium nitride.

Figure 5.4 Single crystal precipitate on the fractured surface of a steel (data acquired from a specimen kindly provided by Dr Monika Jenko, Institute of Metals and Technology, Ljubljana, Slovenia)

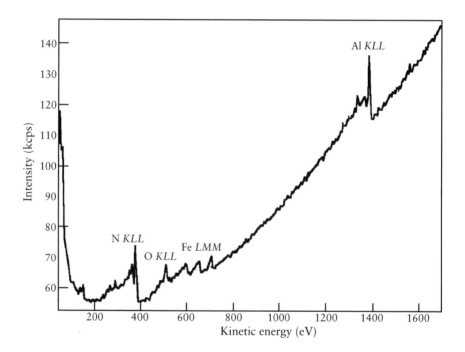

Figure 5.5 Auger spectrum from the single crystal shown in Figure 5.4 (data acquired from a specimen kindly provided by Dr Monika Jenko, Institute of Metals and Technology, Ljubljana, Slovenia)

Figure 5.6 SEM image of a polished section of an alloy susceptible to embrittlement (data acquired from a specimen kindly provided by Dr Monika Jenko, Institute of Metals and Technology, Ljubljana, Slovenia)

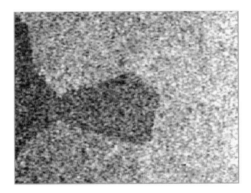

Figure 5.7 Sn Auger map from the polished surface indicating intergranular segregation (data acquired from a specimen kindly provided by Dr Monika Jenko, Institute of Metals and Technology, Ljubljana, Slovenia)

The phenomenon of grain-boundary segregation is illustrated by the scanning electron micrograph of a polished section of an alloy susceptible to such grain-boundary embrittlement shown in Figure 5.6.

The intergranular segregant in this example was tin and as the Sn Auger image, shown in Figure 5.7, shows this element will also segregate to the free (polished) surface following annealing in the UHV chamber of the spectrometer.

The kinetics of segregation can readily be studied by monitoring the segregation to the free surface by holding the specimen at an elevated

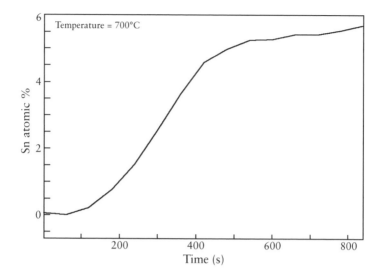

Figure 5.8   Segregant concentration as a function of time as the specimen is held at 700°C (data acquired from a specimen kindly provided by Dr Monika Jenko, Institute of Metals and Technology, Ljubljana, Slovenia)

temperature, within the analysis chamber of the Auger microscope, and plotting segregant concentration as a function of time as illustrated in Figure 5.8. This shows the rapid nature of the diffusion process, as by about 450 s the concentration of Sn measured at the intergranular failure surface was about 5 atomic per cent, equivalent to approximately monolayer coverage of Sn.

5.2.2 Electronic structure of metallic alloys

Although it is Auger electron spectroscopy which has been widely used in metallurgical studies as a result of its superior spatial resolution, XPS is used where information at the chemical or electronic level is required. Alloy design or development has traditionally been carried out on the basis of empirical or systematic methods based on experimental and calculated phase equilibria as presented in the equilibrium phase diagram. Thermodynamic modelling has been applied quite successfully over the last two decades but is essentially phenomenological in nature, and cannot provide adequate information on the subsequent electronic changes (such as charge redistribution) that occur upon alloying. These data are accessible by making so-called first principle calculations but are not easily able to provide phase diagrams of sufficient accuracy, and there is a serious need for experimental evidence regarding the phenomena accompanying alloy formation such as charge transfer and redistribution. This can be achieved by the use of the Auger parameter of the solvent and solute elements of the alloy and the linear potential core model, developed by Thomas and Weightman, to relate the Auger parameter to changes in electronic structure. The description of the model is outside the scope of the text, and the reader is referred to papers cited in the Bibliography, but examples from a Ti–Al–V alloy are given below to indicate the power of the technique.

In order to obtain reliable information from Auger parameter data one must use core-like Auger transitions, i.e., ones where all three electrons involved in the Auger process originate in core-like orbitals rather than the degenerate band structure of the atom. In order to achieve this with metals heavier than magnesium it is necessary to resort to a high-energy X-ray source. Recent work has made use of a $CrK\beta$ source, as this fulfils fourth-order reflections in a conventional $AlK\alpha$ monochromator, which provides a photon energy of 5946.7 eV and a line width of approximately 1.6 eV. The V 1s and V KLL spectra for the Ti–25Al–25V alloy are shown in Figure 5.9, while the Al, Ti and V data relative to the pure metal, and Auger parameters, calculated using the 1s–KLL energies for a range of alloys, are presented in Table 5.1.

It has been suggested that the Auger parameter changes are a measure of the screening efficiency of a system in response to the presence of a localized core hole. Considering that metals are characterized by perfect screening, reduced screening means that the atoms experience a 'less metallic' environment. The Auger parameter shifts (both initial and final state) of the elements of interest between pure metal and alloy in Table 5.1 suggest that the metal with sp valence configuration (i.e., Al) is better screened in the elemental solid than in both the binary and ternary alloys. The difference in the Auger parameter values of Al between the pure element and in the alloys is bigger for the alloy systems with ordering tendency (Ti–Al, Ti–Al–V) than a system with no ordering tendency such as V–Al. Furthermore, the Al Auger parameter is lower for all alloys compared with pure Al, indicating a lower intra-atomic and extra-atomic relaxation or a reduced screening efficiency.

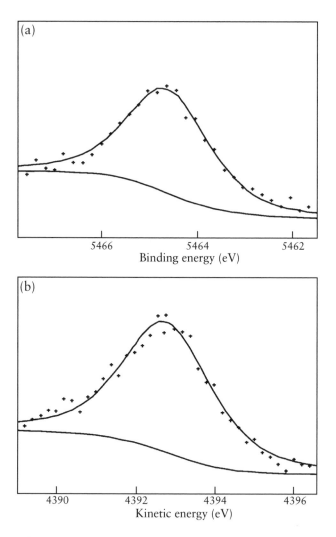

Figure 5.9 The V 1s (a) and V *KLL* (b) spectra for the Ti–25Al–25V alloy (Watts, J. F. *et al.* (2001). *Surf. Interf. Anal.*, **31**, 734–744, Fig. 1, reproduced with permission, 2001. © John Wiley & Sons Limited)

The use of such a high-energy source is a relative luxury but similar qualitative, rather than quantitative, observations can be made with a conventional source. Figure 5.10 shows the Al 2p spectra for a series of Al–Ti alloys compared with pure Al. The chemical shift experienced by the Al in the presence of Ti, compared with the binding energy of the pure Al 2p electrons, is clearly evident.

Table 5.1 Differences in core-level binding energies of Al, V and Ti, and initial and final state Auger parameters, for the alloy relative to the pure metal; the negative sign indicates a decrease in the value for the alloy compared with the pure metal

	Ti–10Al	Ti–20Al	Ti–30Al	Ti–50V	Ti–25V–5Al	Ti–20V–40Al
Aluminium						
$\Delta\alpha$	− 0.9	− 0.7	− 0.7		− 0.7	− 0.6
ΔE_B Al 1s	− 1.1	− 1.0	− 0.9		− 1.0	− 0.8
$\Delta\zeta$	− 2.1	− 1.8	− 2.1		− 2.5	− 2.0
ΔE_B Al 2p	− 0.6	− 0.6	− 0.7			
Titanium						
$\Delta\alpha$	− 0.1	0.0	+ 0.1	− 0.3	− 0.2	− 0.1
ΔE_B Ti 1s	− 0.3	− 0.1	− 0.1	− 0.3	− 0.2	− 0.1
$\Delta\zeta$	− 0.4	− 0.2	− 0.5	− 0.5	− 0.3	− 0.1
Vanadium						
$\Delta\alpha$				+ 0.3	0.0	− 0.1
ΔE_B Ti 1s				+ 0.2	− 0.1	− 0.1
$\Delta\zeta$				+ 0.3	− 0.2	− 0.5
Structure	A3 (hcp)	α2 – DO19	α2 – DO19	A2 (bcc)		B2 (bcc)

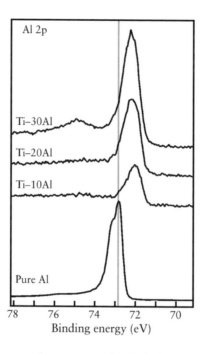

Figure 5.10 Al 2p spectra for a series of Al–Ti alloys compared with pure Al (Watts, J. F. *et al.* (2001). *Surf. Interf. Anal.*, **31**, 734–744, Fig. 3(b), reproduced with permission, 2001. © John Wiley & Sons Limited)

5.2.3 Surface engineering

The modification of a metal surface to provide specific corrosion of tribological properties has been carried out for more than a century but is nowadays recognized as a discipline in its own right: surface engineering. Both AES and XPS are widely used in the analysis of metallic and non-metallic coatings and the analysis complexity varies from the straightforward to the extremely difficult. The former is illustrated by the analysis of an electrodeposited zinc coating on a steel substrate. The coating was known to be several tens of micrometres thick and the specimen was prepared by ball cratering (see Section 4.3.2). Following introduction into the spectrometer, the surface was briefly sputtered to remove adventitious contamination and a linescan recorded from the crater edge to the exposed steel substrate. Using a simple geometrical

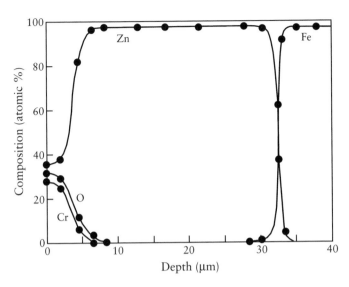

Figure 5.11 Depth profile through a zinc coating on a steel substrate, from a ball-cratering experiment

manipulation, the lateral distance can then be converted to depth and plotted as the depth profile shown in Figure 5.11. The chromium present at the outer surface is the result of the use of a passivation (chromate) treatment following zinc deposition. The zinc deposit is seen to be about 30 μm thick.

Over the last one or two decades, titanium nitride (TiN) and titanium boronitride (Ti–B–N) coatings have become very attractive propositions as wear-resistant coatings. As with many such investigations it is often desirable to correlate the coating composition with performance in some empirical test to establish the optimum coating composition. In the analysis of TiN coatings by AES there is a potential problem as a result of the overlap of the $NKL_{2,3}L_{2,3}$ with the $TiL_3M_{2,3}M_{2,3}$ which is usually studied, as shown in Figure 5.12 for differential Auger spectra of Ti, TiN, and TiB_2.

To overcome this problem it is necessary to resort to using the $TiL_3M_{2,3}M_{4,5}$ at a slightly higher kinetic energy and establishing a calibration curve using specimens of known composition (determined by XPS), which can be plotted relative to the ratio of peak-to-peak intensities of $(NKL_{2,3}L_{2,3} + TiL_3M_{2,3}M_{23})/(TiL_3M_{2,3}M_{4,5})$ as shown in Figure 5.13.

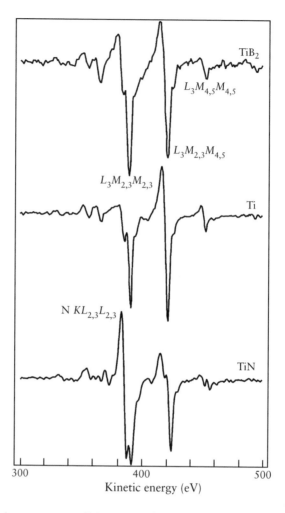

Figure 5.12 Auger spectra of Ti, TiN, and TiB_2 (reprinted with permission from Baker, M. A. *et al.*, "Auger electron spectroscopy/X-ray photoelectron spectroscopy study of Ti–B–N thin films," *Journal of Vacuum Science & Technology A*, **13**(3), 1995, pp. 1633–1638, Fig. 5. Copyright 1995, AVS)

In this case it is helpful to degrade the analyser resolution to negate the effect of fine structure in the Auger spectrum. An alternative approach, which relies on the presence of chemical effects in Auger spectra recorded at high spectral resolution, can be achieved by simply using the $TiL_3M_{2,3}M_{4,5}$ transition alone. This region of the spectrum, recorded in

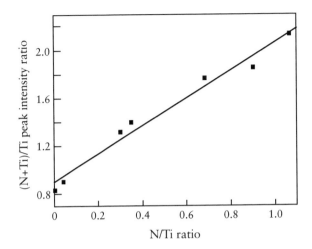

Figure 5.13 Calibration curve to relate the convolution of nitrogen and titanium features to the true N/Ti ratio which has been determined by XPS (reprinted with permission from Baker, M. A. *et al.*, "Auger electron spectroscopy/X-ray photoelectron spectroscopy study of Ti–B–N thin films," *Journal of Vacuum Science & Technology A*, **13**(3), 1995, pp. 1633–1638, Fig. 6. Copyright 1995, AVS)

the direct rather than differential form, is shown in Figure 5.14 for a series of TiN$_x$ coatings, and the evolution of a Ti$L_3M_{2,3}$ hybrid feature at a slightly lower kinetic energy is well resolved in the spectrum.

This hybrid feature is a result of drawing the Ti 3d valence electrons (close to the Fermi Level) towards the N 2p electron (situated below the Fermi Level) and the formation of a hybrid bond. The involvement of the feature in the Auger transition is then seen as a distinct feature in the Auger spectrum, which increases in intensity with increasing nitrogen content. In a similar manner to the previous example it is possible to plot a calibration curve of peak area ratio TiL_3M_{23}hybrid/Ti$L_3M_{23}M_{45}$ which is arguably more accurate than the method described using peak to peak heights of the differential spectra, above. The slight drawback is the need to carry out careful peak fitting and background removal to obtain an accurate estimation of the ratio of components.

In the case of the boronitride it is possible to unravel the complexity of the TiN, TiB$_2$ and BN phases in a complex coating by study of the B 1s and N 1s XPS spectra as shown in Figure 5.15.

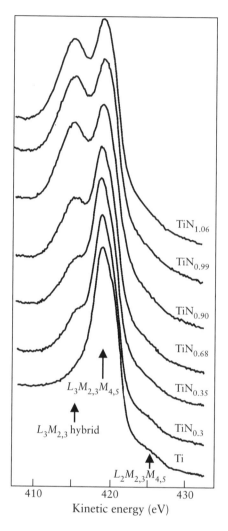

Figure 5.14 $TiL_3M_{2,3}M_{4,5}$ transition for a series of TiN_x coatings (Baker, M. A. *et al.* (2001). *Surf. Interf. Anal.*, **22**, 167–170, Fig. 2, reproduced with permission, 2001. © John Wiley & Sons Limited)

The relative proportions of the three phases are in very good agreement with those predicted by the phase diagram even though the coating in question was deposited under conditions very far from equilibrium! The case for titanium borocarbide coatings is, however, very different. In this coating the phase diagram indicates a three-phase structure of TiB_2, TiC and carbon; the carbon is thought to be in the form of

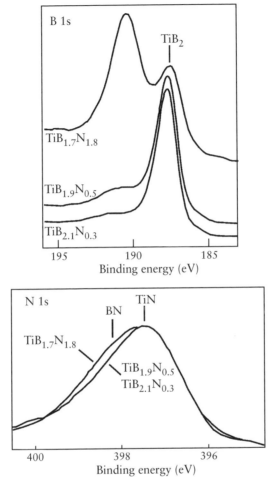

Figure 5.15 B 1s and N 1s XPS spectra from TiN, TiB_2, BN, and TiB_xN_y (reprinted with permission from Baker, M. A. *et al.*, "Combined X-ray photoelectron/Auger electron spectroscopy glancing angle X-ray diffraction/extended X-ray absorption on fine structure investigation of TiB_xN_y coatings," *Journal of Vacuum Science & Technology A*, **15**(2), 1997, pp. 284–291, Fig. 2. Copyright 1997, AVS)

diamond-like carbon (DLC). The C 1s spectra indicate that there is no carbide contribution (Figure 5.16); there is, however, a large component at a binding energy between that of the carbide and pure carbon. This is a result of carbon substitution into the TiB_2 phase instead of forming the two-phase carbide and boride structure.

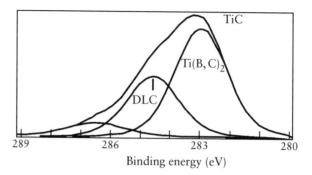

Figure 5.16 The C 1s spectra from titanium borocarbide (Baker, M. A. *et al.* *ECASIA'97 Proceedings*, p. 1128, Fig. 3, Olefford, Nyborg and Briggs (eds). 1997. © John Wiley & Sons Limited, reproduced with permission)

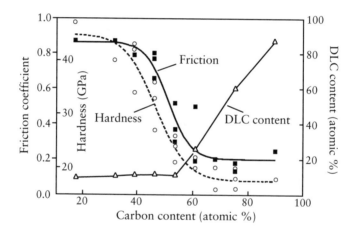

Figure 5.17 Correlation of total carbon and DLC content of a series of titanium borocarbide fibres with friction and hardness parameters (Baker, M. A. *et al.* *ECASIA'97 Proceedings*, p. 1130, Fig. 5, Olefford, Nyborg and Briggs (eds). 1997. © John Wiley & Sons Limited, reproduced with permission)

At lower carbon concentrations this occurs in preference to the formation of the DLC phase. The concentration of DLC as a function of total carbon has been quantified by XPS and this indicates very well why such high levels of carbon (>50 atomic per cent) are required before the expected reduction in the coefficient of friction and hardness are observed (Figure 5.17).

These two examples serve to indicate the power of XPS to identify and quantify surface phases, it is then possible to relate such observations back to the phase diagram to establish whether equilibrium or metastable phases exist in the coating.

5.3 Corrosion Science

There are two areas within corrosion science in which electron spectroscopy has had a dramatic impact: the interaction of a metal surface with its environment, perhaps to form a passivating overlayer, and the breakdown of the surface film by a localized phenomenon such as pitting. The former is readily studied by XPS, where the ability to separate the spectrum of the underlying metal from that of its oxide film enables the definitive identification of passivating films on alloys. It also enables the thickness of very thin films to be estimated using the Beer–Lambert equation of Chapter 1.

However, although XPS provides valuable information concerning the composition and perhaps growth kinetics of such films, their breakdown leading to major environmental degradation of the metal is a localized phenomenon requiring high spatial resolution surface analytical methods, i.e., sub-micron scanning Auger microscopy. The provision of compositional depth profiles of corrosion films is a standard requirement and may be achieved by using argon ion bombardment in conjunction with either AES or XPS. In the case of very thin films (<4 nm) angle resolved XPS can be very informative.

Although in some cases the deconvolution of the metallic and cationic spectral information is straightforward, there are cases of great importance to the corrosion scientist where a major research effort is required to unravel the complexities of two or more valence states along with loss features present in the spectrum. The transition metals in general have presented difficulties. In the case of iron, the need is to be able to distinguish Fe(II) from Fe(III) confidently, as shown in Figure 5.18.

The two spectra indicate that there is a small difference in binding energy between the $2p_{3/2}$ position of the chemical states. However, it is the position of the shake-up satellites in the valley between the $2p_{3/2}$ and $2p_{1/2}$ components that give the most reliable information. In the case of

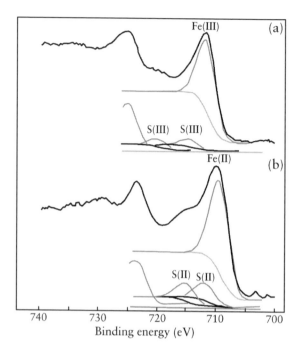

Figure 5.18 Curve fitting of iron $2p_{3/2}$ spectra of (a) Fe(III) and (b) Fe(II) (the satellites (S) provide a valuable aid to assignment of chemical state)

copper, the shake-up satellites provide a clear route to the identification of copper (II), as seen in Figure 3.6, but the separation of metallic copper and the monovalent ion requires the use of the X-ray induced Auger transition.

In the case of the Fe(II) state, these features give rise to a broadening on the lower binding-energy side of the valley between the Fe $2p_{3/2}$ and Fe $2p_{1/2}$ components (Figure 5.18(a)), for Fe(III) on the higher-energy side (Figure 5.18(b)). The curve fitted spectra of Figure 5.18 show the relative positions and intensities of the satellites. In the case of a mixed phase, all these components, together with Fe(0) (if the film is very thin) must be considered, preferably for both 2p peaks as a check for self-consistency. The backgrounds of the individual singlets will also vary with depth distribution. It is only by using sophisticated computer curve-fitting routines in combination with extensive knowledge of peak shape and relative intensities, that this has become possible. This type of

information is well documented for all elements of interest to the corrosion scientist although, in some cases, it is necessary to consider the X-ray
induced Auger peaks in the XPS spectrum as well. These are particularly
informative for magnesium, copper, and zinc, where the photoelectron
spectra alone do not provide unambiguous chemical state information.

A little used, but nevertheless valuable, method in which surface analysis can assist in the determination of electrochemical history, is by the
monitoring of cations or anions absorbed from aqueous solutions. For
instance, if a metal electrode is polarized cathodically in $MgCl_2$ solution, it will preferentially adsorb cations on the metal surface, seen in
the electron spectrum (XPS or AES) as an excess of magnesium to
chlorine. If the electrode has been polarized anodically the reverse is
true. This approach is useful in establishing if a pit, or other corrosion
site is active or benign, it can also be used to assess the potential
distribution around an active pit. An experiment of the latter type is
described in Figure 5.19. The broken line and right-hand axis indicates
the anion/cation ratio determined on large electrodes as a function of

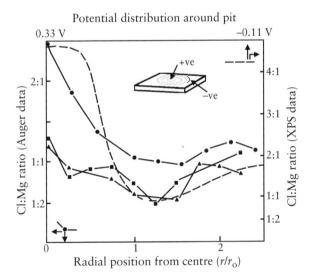

Figure 5.19 Electrode potential around a pit: the dashed line represents the
predicted potential distribution obtained from large area (XPS) analysis and
consideration of pit geometry (upper and right-hand side axes); the data points
represent Auger analyses made on very fine pits (lower and left-hand side axes)

electrode potential for a constant charge passed, plotted against the predicted potential distribution for small pits. Microanalysis of three pits of about 2 μm in diameter by Auger spectroscopy allows the electrode potential distribution around a pit to be determined experimentally, and shows excellent agreement with that predicted by the broken line (the combination of XPS and theory), indicating an active anodic centre to the pit surrounded by a cathodic halo as illustrated in the schematic diagram of Figure 5.19.

Such pitting may be related to microstructural features of the alloy, and the addition of X-ray analysis facilities to a scanning Auger microscope enables features such as inclusions to be identified and imaged at the same time as the surface phases. In Figure 5.20 a combination of Auger and X-ray maps indicate that a pit, identified in the SEM image is shown to be active (by the concentration of chlorine in the surface analysis), and associated with a CuS inclusion group, identified by EDX. Figure 5.20 also summarizes the other information of value in

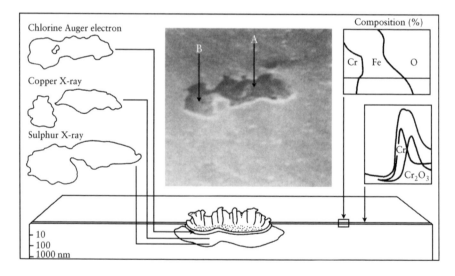

Figure 5.20 A combination of analytical methods has been used to define the corrosion processes occurring around the inclusion group identified as sulphide containing by EDX (middle left): the passive film is characterized by Auger depth profiling and XPS (upper right); the chlorine Auger map indicates the active area which corresponds to the pit observed in the SEM image (which is 6 μm wide)

corrosion studies, the chemical state information from XPS, and the compositional change within the very thin passive film obtained by sputter depth profiling.

The versatility of a scanning Auger microscope fitted with an energy dispersive X-ray detector has been illustrated by studies of the initiation of pitting corrosion in a stainless steel at oxide and manganese sulphide inclusions. A system of this type is illustrated in the next Chapter, (Figure 6.3, p. 169). In order to follow the corrosion process as a function of time it is necessary to re-establish the specimen repeatedly in the Auger system for analysis, following exposure to the corrosive environment. This is best achieved by the judicious use of microhardness indents around the inclusion group of interest. Although inclusions in steels will have good contrast in optical microscopy this is not the case with electron microscopy and, until corrosion features are evident in the secondary electron image, relocation of the specimen is a very uncertain process. Hence the need for physical markings such as the microhardness indents.

A typical oxide inclusion is shown in Figure 5.21(a) following 1 day of exposure to acidified sodium chloride solution, whilst Figure 5.21(b) is the same inclusion after 63 days exposure. A complementary set of SAM and EDX images, taken after 30 days exposure is presented in Figure 5.22.

The micrographs of Figure 5.21 indicate that pitting has initiated, as expected, at the oxide inclusion/metal boundary, and the X-ray images of Figure 5.22 show that the inclusion is a mixture of Mn/Ti/Al oxides. The SEM images show the presence of corrosion deposits adjacent to the inclusion, the surfaces of which are enriched in oxygen, chlorine, silicon, titanium, and managanese, as indicated by the Auger images. Clearly there has been a reduction of pH within the crevice which has developed at the oxide/metal interface which has lead to the partial dissolution of the inclusion and the deposition of titanium and manganese ions on the adjacent regions. These zones have also been decorated by silicon which is thought to be the result of the dissolution of a soluble silicate which is a minor component of the oxide inclusion.

A similar approach has been used to study the very early (after 10 s exposure to saline solution) dissolution of manganese sulphide inclusions in steel. The model of Figure 5.23, indicates the reactions and transport processes at, and around, a pit, as a function of time, associated with such inclusions and was deduced from Auger analyses.

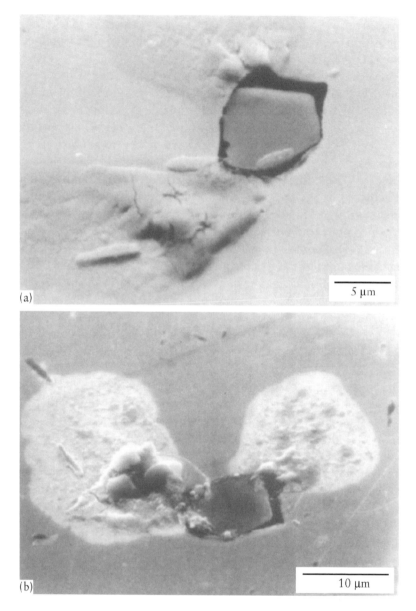

Figure 5.21 A typical oxide inclusion (a) following 1 day of exposure to acidified sodium chloride solution, and (b) after 63 days exposure (reprinted from Baker, M. A. and Castle, J. E. The initiation and pitting corrosion of stainless steels at oxide inclusions. *Corrosion Science*, **33**, pp. 1295–1312, Fig. 2. Copyright 1992, with permission from Elsevier Science)

The benefits of this approach to the investigation of corrosion phenomena are that reactants and products associated with the chemical reactions can be identified *in situ*, and reaction mechanisms proposed on the basis of analytical chemistry results rather than inferred from electrochemical measurements and morphological observations.

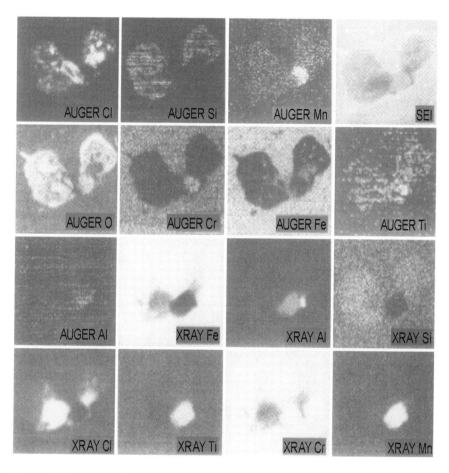

Figure 5.22 A complementary set of SAM and EDX images, taken after 30 days exposure to acidified sodium chloride solution (reprinted from Baker, M. A. and Castle, J. E. The initiation and pitting corrosion of stainless steels at oxide inclusions. *Corrosion Science*, **33**, pp. 1295–1312, Fig. 3. Copyright 1992, with permission from Elsevier Science)

(a)

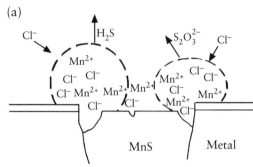

Initial stages of MnS dissolution

(a) Anodic nature of dissolving inclusion and loss of S-containing anions from electrolytic system (formation of H_2S, elemental S as corrosion products) attracts a high local concentration of Cl^- to the site.

(b) Largest concentration of Mn^{2+} and Cl^- accumulating in the bulk solution above the small cavity.

(c) Exposed bare metal repassivates.

(b)

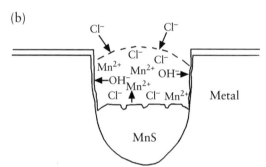

Partial MnS dissolution

(a) Futher dissolution causes an increasing Mn^{2+} and Cl^- concentration within the cavity.

(b) Uni-directional diffusion restricts the transport of corrosion products away from the cavity.

(c) Exposed cavity wall repassivates.

(c)

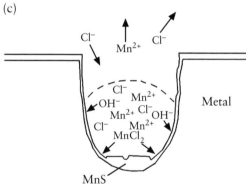

Further MnS dissolution and stabilization of pit growth

(a) Critical concentration of Mn^{2+} and Cl^- attained for precipitation of $MnCl_2$ salt film.

(b) The salt film prevents repassivation and provides conditions favouring stabilized metallic corrosion.

Figure 5.23 Model of the early dissolution of manganese sulphide inclusions in steel (reprinted from Baker, M. A. and Castle, J. E. The initiation and pitting corrosion at MnS inclusions. *Corrosion Science*, **34**, pp. 667–682, Fig. 10. Copyright 1993, with permission from Elsevier Science)

5.4 Ceramics and Catalysis

In the field of ceramics it has been XPS that has proved the more useful of the two techniques, with problems of sample charging limiting the number of investigations carried out by AES. However, the last few years have seen a widening of the use of Auger spectroscopy in the fields of catalysis and mineralogy and one can be sure that this trend will continue. In this section we shall consider the role which electron spectroscopy has to play in the analysis of catalysis samples and naturally occurring minerals.

In the application of XPS to catalysis studies, there appear to be three areas of endeavour.

1. The furthering of the basic science of heterogeneous catalysis has relied greatly on the pure surface science approach, i.e., the preparation of metal or inorganic single crystals with a pre-defined crystal orientation which is then exposed to very small quantities of reactant(s) in the gas phase. In this manner, the reaction occurring on the crystal surface can be followed in a stepwise pattern, the modification of substrate or adsorbate being apparent from the electron spectrum. Such experiments are often carried out in conjunction with low-energy electron diffraction (LEED) which yields information concerning surface crystallography.

2. The activity of a supported catalyst is frequently a function of the level of dispersion of the metal or oxide on the support medium. The size of such supported crystallites can sometimes be estimated from the intensity of the appropriate XPS peak ratio. However, this does require assumptions regarding particle shape and the most rewarding studies appear to be those which combine XPS data with a TEM study.

3. The area in which surface analysis has made the most spectacular impact is in the identification of catalyst poisons, and other trouble-shooting investigations.

An example of the use of AES in comparing fresh and spent catalyst is presented in Figure 5.24. The two spectra were obtained from a

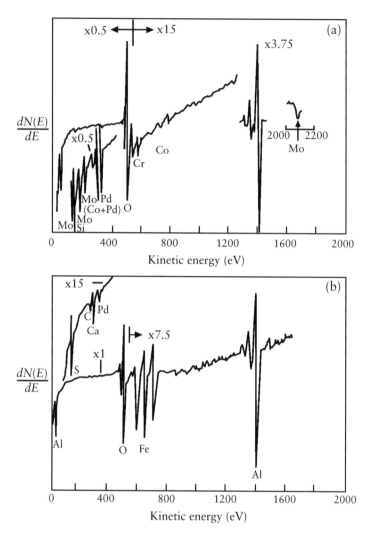

Figure 5.24 AES of (a) fresh, and (b) spent alumina-supported 0.5 per cent Pd catalyst (figure from Bhasin, M. M., *Journal of Catalysis*, **38**, 218–222, Copyright 1975, Elsevier Science (USA), reproduced with permission from the publisher)

0.5 per cent Pd catalyst supported on Al_2O_3 with chromium and molybdenum additions as promoters. Comparison of the Auger spectra from the two samples indicates that de-activation is associated with a large concentration of iron attenuating the Al, Pd, Cr, and Mo signals present on the clean surface. Thus, the poor performance of this material could be associated with an iron contaminant, probably emanating from steel pipework or the reaction vessel, masking the highly active palladium atoms as well as the promoter atoms, and greatly reducing the catalytic activity of the material.

The surface analysis of naturally occurring minerals, although straightforward in principle, provides many problems in practice. Whilst the provision of an elemental surface analysis by XPS is straightforward, extracting the required level of chemical information can be difficult. There are two problems involved. First, electrostatic charging often means that the confidence level on the charge-corrected peak position is greater than the spectral window containing the chemical information! Second, the peaks often have a very poor photoelectron cross-section (e.g., Si 2p and Al 2p). The methods by which these problems have been overcome are closely related to each other and have been pioneered by Castle's Group at the University of Surrey UK.

The problem of sample charging may be overcome by reporting the separation of two peaks rather than the absolute binding energy of a photoelectron transmission. In some elements the Auger chemical shift is equal to, and sometimes greater than, the XPS chemical shift, and the potential exists for extracting chemical information via the Auger parameter (α) defined in Chapter 3, as:

$$\alpha = E_K + E_B$$

where E_B is the binding energy of the XPS peak (e.g., 1s or 2p) and E_K is the kinetic energy of the attendant Auger peak (e.g., $KL_{2,3}L_{2,3}$). In the case of aluminium and silicon the $KL_{2,3}L_{2,3}$ Auger transition is not directly accessible although the bremsstrahlung radiation from a conventional X-ray gun is able to eject sufficient Si 1s electrons to produce a measurable $SiKL_{2,3}L_{2,3}$ peak. The silicon Auger parameter calculated in this way is independent of sample charging but strongly dependent on both molecular and crystalline structure.

There still exists the problem of poor photoelectron cross-section of the Al 2p and Si 2p levels and the only way to overcome this is to use a higher-energy X-ray anode able to excite 1s core levels of these elements. Various possibilities exist including ZrLα and TiKα, but the ones which represent the best combination of sensitivity and resolution are monochromated AgLα and CrKβ. Using such a source it is possible to record 1s-KLL Auger parameters; a spectrum obtained with a

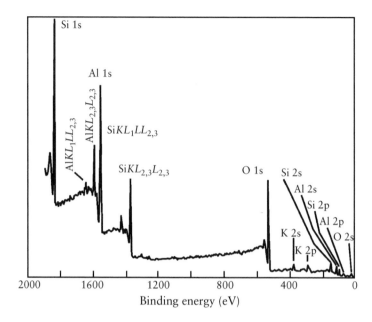

Figure 5.25 XPS survey spectrum of muscovite mica recorded with monochromatic AgLα radiation

monochromated AgLα high-energy source of this type is shown in Figure 5.25.

One final application of XPS in the analysis of ceramics is that of photoelectron forward scattering, also referred to as X-ray photoelectron diffraction (XPD). If the specimen under consideration is a single crystal, photoelectrons emitted from sub-surface planes will be scattered by surrounding atoms giving rise to an angular distribution of the emitting site concerned.

This type of angular modulation is quite different from that utilized in the non-destructive depth profiling of materials described in Chapter 4. Indeed for successful forward scattering, the crystal surface should be extremely clean, preferably prepared *in vacuo* by, for instance, cleaving. This procedure has been used with a high degree of success for the investigation of complex minerals, and also semiconductors such as gallium arsenide. It would seem that such a method offers much promise as a way of ensuring the epitaxial orientation of very thin layers.

5.5 Microelectronics and Semiconductor Materials

The trends in the development of semiconductor devices continue to be towards both increasing transistor density and higher operating frequencies. The consequences of this are that individual structures on the chip are becoming smaller and thinner and the size of defects which cause chip failure (critical defects) are also becoming smaller. The fact that these devices are built at or near the surface of a silicon wafer means that surface analysis techniques are commonly used in the semiconductor industry.

The spatial resolution available from Auger electron spectroscopy makes this a useful tool for the analysis of device structures. The depth profiling capabilities of the technique are also used to assess the quality of the layered structures which form part of modern semiconductor devices. XPS is also used for its ability to determine the chemical states of the, sometimes complex, materials which are produced.

The trend towards thinner layers means that dielectric films are becoming amenable to study using ARXPS. The need for XPS is expected to increase when silicon dioxide is eventually replaced by more exotic materials having a much higher dielectric constant. With these layers, it will become important to ascertain the chemistry not only of the layer but also of the interface between the layer and the silicon or silicon dioxide substrate. ARXPS will be able to do this without the need to sputter the material and risk the alteration of the chemical states present.

5.5.1 Mapping semiconductor devices using AES

Figure 5.26 shows the maps which can be obtained from the cross-section of a semiconductor device. In this case, the features present in, for example, the titanium map are about 50 nm wide, illustrating the ability of Auger to provide analyses at very high resolution. In addition, the Al map clearly shows discontinuities which could cause device failure.

Auger mapping is capable of more than elemental analysis. If the instrument has reasonable energy resolution then chemical-state maps

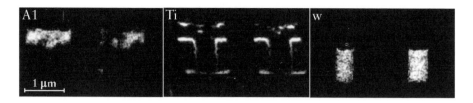

Figure 5.26   Auger maps from the cross-section of a semiconductor device

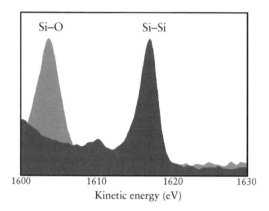

Figure 5.27   Auger spectra from elemental silicon and oxidized silicon showing the chemical shift which occurs in the oxide

can be produced. When the separation of Auger peaks is large, as is the case with elemental silicon and its oxide (Figure 5.27) then mapping the chemical states is straightforward.

Figure 5.28 shows another Auger map from a sectioned piece of silicon. In this case, two chemical states of silicon were mapped revealing an oxide layer only 15 nm thick.

Using even better energy resolution, mapping of dopant types becomes possible. Figure 5.29 shows that there is a small energy difference between the Si *KLL* peaks from n type and p type silicon. A similar phenomenon is also observed in XPS.

The concentration of the dopants in these materials is extremely small, far below the detection limits of Auger electron spectroscopy. Nevertheless, the energy shift is sufficient to allow maps to be produced.

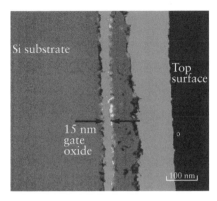

Figure 5.28 The chemical states of silicon (elemental and oxide) can be mapped with high spatial resolution using AES

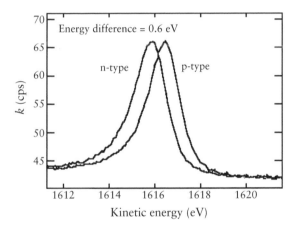

Figure 5.29 Spectra from n-type and p-type silicon show a small shift in peak position (0.6 eV)

As an example, Figure 5.30 shows a map derived from the cross-section of a silicon sample implanted with phosphorus to produce n-type silicon.

An alternative method for the analysis of this type of sample is to produce a peak shift profile from an Auger line scan through the doped layer. Such a line scan is shown in Figure 5.31.

Figure 5.30   Overlay of maps derived from n-type (left) and p-type (right) silicon

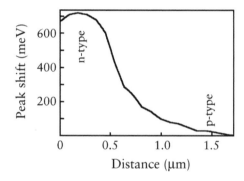

Figure 5.31   An Auger line scan through a p–n junction

5.5.2 Depth profiling of semiconductor materials

Sputter depth profiling using Auger electron spectroscopy is frequently applied to semiconductor materials. The very flat interfaces present in many of the sample types means that extremely good depth resolution can be achieved if all of the instrumental parameters are carefully chosen, as discussed in Chapter 4.

Elemental profiles are performed to ascertain layer and interface purity, layer thickness and migration or diffusion of material from one layer to another. However, chemical-state information is frequently required

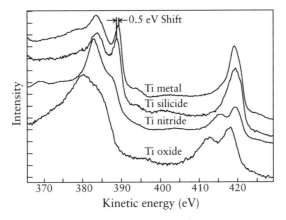

Figure 5.32 Auger spectra from elemental titanium and some of its compounds

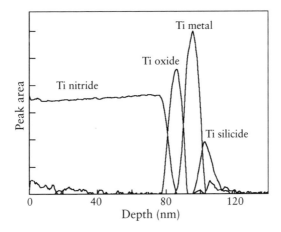

Figure 5.33 Depth profile of titanium nitride/titanium oxide/titanium on silicon (note the formation of the silicide at the silicon interface)

and can be provided by profiling using Auger electron spectroscopy. Figure 5.32 shows Auger spectra from elemental titanium and some of its compounds. A profile data set containing titanium spectra can then be fitted with these standard spectra using a least squares method. Figure 5.33 shows the results of this process from a multilayer material on a silicon substrate. For clarity, only the titanium species are shown.

The method of least-squares fitting is sufficiently powerful to distinguish titanium silicide and titanium metal when the spectra are very similar; the silicide spectrum is shifted with respect to the metal spectrum by only 0.5 eV.

5.5.3 Ultra-thin layers studied by ARXPS

Modern dielectric layers, which are used in the gate region of transistors, need to be extremely thin to allow rapid and efficient switching of the transistor without the need for high voltages. The thickness of these materials is becoming similar to the information depth of X-ray photoelectron spectroscopy.

Figure 5.34 shows a set of data from layers of silicon oxide on silicon nitride on silicon. Each of the spectra was taken at a different emission angle and clear changes can be seen in the spectra as a function of angle. The least-squares fitting techniques can be applied to the data, using silicon, silicon dioxide and silicon nitride spectra as standards and the

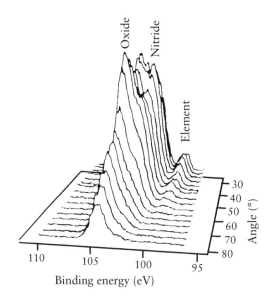

Figure 5.34   An ARXPS data set from silicon dioxide on silicon nitride on silicon

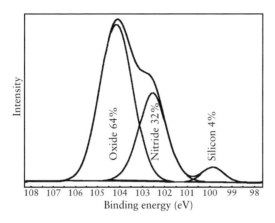

Figure 5.35 Quantification of one of the spectra shown in Figure 5.34

relative quantity of the three components calculated for each of the experimental spectra; this is shown in Figure 5.35 for one of the spectra in Figure 5.34. This allows the layer thicknesses to be calculated using methods described in Chapter 4. In this case, the oxide layer was found to be 4 nm thick and the nitride layer 5 nm.

5.6 Polymeric Materials

Since its inception as a commercial technique some three decades ago, XPS has been widely used as a technique for the surface chemical analysis of polymers. Its acceptance in this role is mainly due to Clark's Group at the University of Durham, UK, who were responsible for much of the careful characterization work in the early days of its development. The last decade has seen considerable advances in the XPS analysis of polymers and this can be traced back to the systematic study of a wide range of polymers using a new class of spectrometer featuring a monochromatic AlKα source and a high transmission electron energy analyser. The use of such a high-resolution source was not new, indeed at that time monochromatic sources had been available for more than a decade, but acquisition times had always been extremely long, long enough to preclude the use of monochromators for routine analysis. The development of high-sensitivity analysers, and high-power rotating anode X-ray sources, within Siegbahn's Group at Uppsala, Sweden, led to a commercial product. The launch instrument was installed in ICI's Wilton Laboratories, UK, in 1989. Graham Beamson and David Briggs at ICI used this spectrometer to record high-resolution spectra of over 100 homopolymers. These were subsequently published as the *High Resolution XPS of Organic Polymers*, although now out of print the spectra are available in electronic form from SurfaceSpectra Ltd, Manchester (www.surfacespectra.com/xps)

The influence of the prototype spectrometer, and the seminal handbook that sprang from the polymer XPS data acquired on it, cannot be overemphasized: the routine use of a high-resolution X-ray source provided a much greater level of information than had previously been obtainable. The issue of charge compensation using an electron flood gun for the XPS analysis of polymers had also been established at a higher level of precision. In a very short time all the major manufacturers of XPS systems were offering high-resolution XPS systems with performance approaching, and eventually surpassing, that of the ESCA300. One area of fertile development has been in the design of charge compensation systems. Thus, although good quality XPS can be carried out on polymers using twin anode (non-monochromatic) sources, for the best quality analysis the use of a monochromatic source is necessary.

As all organic polymers contain substantial quantities of carbon, it is the chemical shift of the carbon 1s electrons which predominates the interpretation of XPS data from these materials. Figure 5.36 shows the C 1s spectrum of poly(methyl methcrylate) recorded using a monochromatic AlKα source. In order to achieve satisfactory peak fitting of the experimental spectrum it is necessary to use four singlets. These peaks correspond to aliphatic carbon at a binding energy of 285 eV (a useful internal standard in polymer analysis), carboxyl carbon at a separation of approximately 4.2 eV from the C–C/C–H peak, and ether-like carbon at a distance of about 1.8 eV (as indicated on the structural formula of this polymer, which is also included in Figure 5.36).

The final component at a separation of 0.7–0.8 eV is due to a secondary chemical shift, which is the effect of the carboxyl group on the unsubstituted carbon atom in the $C–CO_2R$ structure. Such secondary shifts (also known as nearest neighbour effects) have only been reported in the literature relatively recently and it is clear that their identification results from improvements in peak-fitting methods as well as the

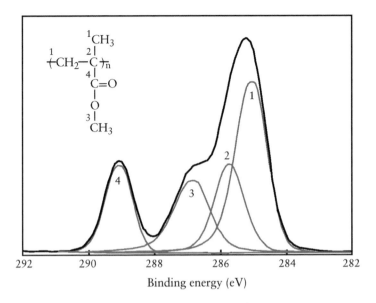

Figure 5.36 C 1s spectrum of poly(methyl methacrylate) recorded with a monochromatic AlKα source; the structural formula of this polymer is also shown and the carbon atoms identified to indicate their contribution to the spectra (*High resolution XPS of organic polymers*, G. Beamson and D. Briggs, p. 119 C 1s. Copyright 1992. © John Wiley & Sons Limited. Reproduced with permission)

widespread use of monochromatic X-ray sources. Peak fitting is now invariably carried out by computer methods but much of the early work was achieved using much less sophisticated techniques. In such instances, the secondary shift is accounted for by merely making the methyl carbon peak slightly wider and more intense. Unless the carboxyl peak is fairly strong, the secondary shift is easily lost in the vagaries of the peak fitting exercise when using MgKα radiation, although the situation is much more secure in the higher-resolution spectrum, obtained with a monochromatic source.

By careful use of XPS, it is possible to differentiate between aliphatic and aromatic carbons, there are two possible ways in which this can be done. In the case of high-resolution XPS a small, but significant, chemical shift of about − 0.5 eV occurs in the aromatic species relative to aliphatic unfunctionalized carbon atoms. The spectrum of Figure 5.37 is taken from polystyrene and the shake-up satellite resulting from the

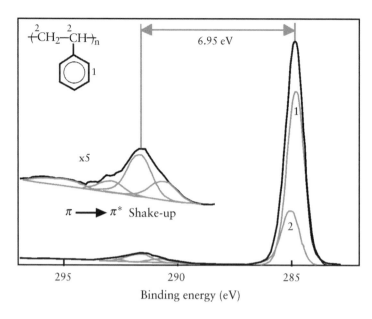

Figure 5.37 C 1s spectrum of polystyrene showing the $\pi \to \pi^*$ shake-up satellite (*High resolution XPS of organic polymers*, G. Beamson and D. Briggs, p. 73 C 1s. Copyright 1992. © John Wiley & Sons Limited. Reproduced with permission)

$\pi \to \pi^*$ transition in the phenyl ring, which accompanies photoemission, can be seen as a discrete feature some 7 eV from the main peak. The intensity of the satellite as a function of the main photoelectron peak remains constant at about 10 per cent although slight changes occur depending on the structure of the polymer involved. This feature provides a quantitative way in which the surface concentration of phenyl groups, following a particular treatment method, may be estimated. It also provides a means of estimating surface modification brought about by ring opening reactions.

In addition to the use of C 1s and other core levels, the valence band of the XPS spectrum can be particularly useful for polymers. Once again, the monochromatic source is recommended but this region can often provide very useful, qualitative, information. The XPS valence band spectra of poly(ethylene) and poly(propylene) are illustrated in Figure 5.38 and although the C 2p regions are very similar clear differences in the C 2s region of the spectrum are seen. For poly(ethylene) the region 12–25 eV is composed of two readily identifiable features, whilst

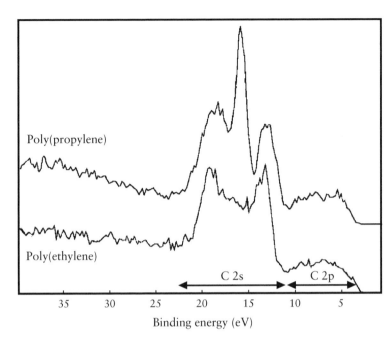

Figure 5.38   Valence band regions of poly(ethylene) and poly(propylene)

for poly(propylene) there are three clear components in this region of the spectrum. The C 1s core levels are very similar although both show slight broadening as a result of vibrational effects.

In many instances valence band XPS spectra of polymers are used in a 'fingerprinting' manner to differentiate between similar systems but it is also possible to compare the experimental spectra with cluster calculations or another numerical approach. For practical surface analysis, the former comparison is often sufficient, as one should not underestimate the complexity or time-consuming nature of adopting the second approach!

Although XPS has been widely used to study thermoplastic polymers such as poly(methyl methacrylate) and poly(styrene) its use to study crosslinked systems is not so widespread. Such materials are widely used as organic coatings and adhesives which are invariably sophisticated formulations of many components to provide the required mechanical, thermal, process, and aesthetic properties. The many organic components ensure that the resultant C 1s spectrum is extremely complex and in order to resolve all the components in the formulation it is necessary

Epoxy reactive resin

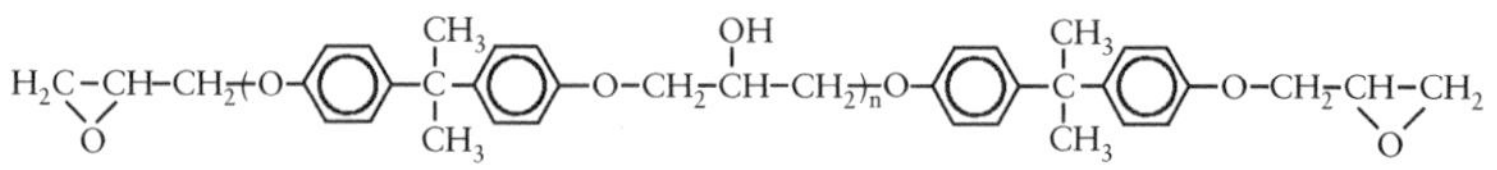

Urea formaldehyde crosslinking agent

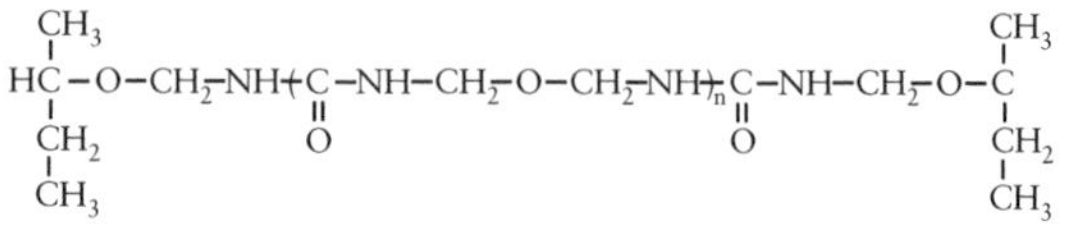

Acrylic flow agent

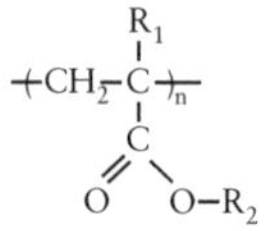

where R_1 and R_2 are H or alkyl pendent chains

Figure 5.39   Components of a thermally-cured urea formaldehyde/epoxy coating

to carry out XPS at the highest possible resolution. The following
example indicates the complex nature of C 1s spectra from even very
straightforward organic coatings. A thermally-cured urea formaldehyde/
epoxy coating, containing the components indicated in Figure 5.39, was
analysed by XPS using monochromatic AlKα radiation; formulations
with and without an acrylic flow agent were examined. The resultant
spectrum of the formulation containing the flow agent can be peak fitted
to identify the presence of 11 separate components (Figure 5.40). Eight
of the components arise from the urea formaldehyde/epoxy coating,
but three uniquely identify the flow agent. The peaks characteristic of
the flow agent are assigned to carbon atoms in β-position from a
carboxyl group (C–COO, 285.49 eV), an ester carbon component
(C–O–C=O, 286.67 eV) and carbon species involved in carboxyl
groups (O=C–O, 289.16 eV) respectively.

   To investigate the near surface depth distribution of the various
elements and carbon functionalities angle resolved XPS was carried
out on the coating with and without the flow agent. The C 1s, N 1s
and O 1s angle resolved data sets, acquired on a parallel angle resolving
spectrometer, for the coating without the flow agent are presented in
Figure 5.41.

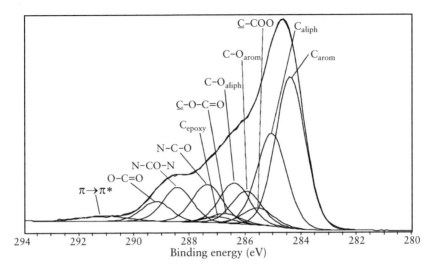

Figure 5.40 XPS spectrum from a thermally-cured urea formaldehyde/epoxy coating (*Surf. Interf. Anal.*, **33**(10), Fig. 3, Perruchot, C. *et al.* Copyright 2002. © John Wiley & Sons Limited. Reproduced with permission)

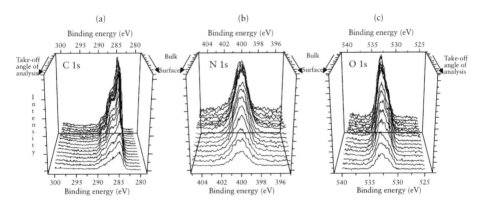

Figure 5.41 ARXPS measurements on the thermally-cured urea formaldehyde/ epoxy coating with a flow agent (*Surf. Interf. Anal.*, **33**(10), Fig. 6, Perruchot, C. *et al.* Copyright 2002. © John Wiley & Sons Limited. Reproduced with permission)

In all but very special cases, depth profiling of polymers by ion sputtering is impractical because of gross sample degradation and the usual methods adopted are angle resolved XPS or multi-photon investigations. An angle resolved data set does not in itself yield a compositional depth profile and a suitable algorithm is required to reconstruct

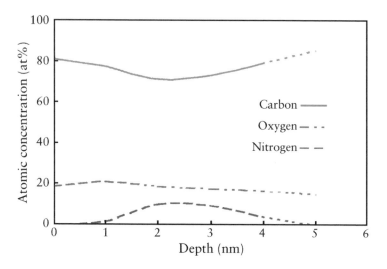

Figure 5.42 Reconstructed depth profile from angle resolved XPS data of a urea formaldehyde/epoxy coating showing the surface segregation of the acrylic component (*Surf. Interf. Anal.*, **33**(10), Fig. 8(b), Perruchot, C. *et al.* Copyright 2002. © John Wiley & Sons Limited. Reproduced with permission)

the depth profile from quantitative XPS results. One such method of achieving this end is the routine ARCtick available from the National Physical Laboratory in the UK (www.npl.co.uk). This routine has been used on an angle resolved XPS data set from the urea formaldehyde/epoxy plus flow agent coating discussed above, and the resulting depth profile is shown in Figure 5.42. As expected the nitrogen concentration, indicative of the coating itself, is depressed about 1 nm below the surface as a result of the surface segregation of the acrylic flow aid, which forms a layer about 1 nm thick at the surface.

The alternative method of depth profiling, changing the energy of the X-ray photons, is potentially a very elegant way with which to probe different depths. In practice, however, it is rather restrictive and most spectroscopists will be limited to a combination of AlKα and MgKα which gives a depth selectivity of about 1 nm on the C 1s line.

There are certain instances where the concentration of particular functional groups at a polymer surface are low and peak fitting of the carbon 1s spectrum can become a rather uncertain process. In these instances, a method known as chemical derivatization can then be used.

In essence the surface functional groups are reacted with a liquid or gas phase reagent which tags them with an atom or ion which is readily determined by XPS. The concentration of the characteristic element is then directly proportional to the concentration of the functional group involved. Many such derivatization reactions exist using both organic and inorganic reagents, the most successful being those where the chemical tag has a high cross-section in XPS such as barium, thallium, or silver. It should be pointed out, however, that with the advent of high-resolution XPS using a monochromatic source the popularity of this approach appears to have declined.

The vast majority of polymer science investigations making use of XPS report the changes that have occurred as a result of surface modification, either by process treatment (to improve surface properties such as wettability), or naturally occurring phenomena (such as the weathering of paint films). Recent work, however, has concentrated on the identification of surface segregation and depletion phenomena in investigations involving multi-component systems.

5.7 Adhesion Science

There are three distinct areas in which surface analysis has made major contributions to the science and technology of adhesion: the analysis of surfaces prior to the application of the coating or adhesive, and the subsequent correlation of adhesion with surface cleanliness; the investigation of the substrate to polymer bond; and the exact definition of the locus of failure following bond failure. Each of these areas will now be considered in turn.

The cleanliness of a metal substrate prior to its contact with a polymer adhesive or coating, is readily assessed by XPS or AES. Although gross levels of contamination such as oils from mechanical working or temporary corrosion protection can easily be detected by other methods, it is only surface sensitive techniques which can assess the efficacy of a cleaning method. As an example, the spectra of Figure 5.43 are taken from a steel sheet that has been prepared by alkaline cleaning (Figure 5.43(a)) and emery abrasion (Figure 5.43(b)). The level of carbonaceous contamination is considerably higher on the solution-cleaned surface indicating that abrasive cleaning produces the

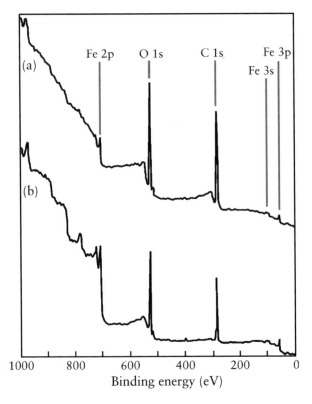

Figure 5.43   Surface cleanliness of a steel sheet: an alkali cleaning process (a) leaves substantially more contamination than an abrasive cleaning process (b)

better quality surface from a chemical point of view. Studies some 20 years ago by the American automobile industry established unequivocally that high levels of surface carbon were a contributory factor to poor durability of painted steel from certain manufacturers. Another frequently cited example is that of aluminium-magnesium alloys which, if heat-treated incorrectly, develop a friable surface film of magnesium oxide. Adhesion of a paint film or adhesive to such substrates is very poor and failure occurs rapidly. XPS or AES are able to identify such a layer and these techniques can be used in a diagnostic manner prior to bonding or painting to ensure that adhesion will be of the required standard. Contamination may also arise from a variety of external sources – sub-monolayer coverage of aluminium surfaces by fluoro-carbons has been shown to produce a drastic reduction in adhesive bond strength.

Once the bond has been made, the task of examining the interfacial chemistry is extremely difficult. Various methods of approaching the interface have been developed, involving the removal of the polymer in a suitable solvent, or the dissolution of the iron substrate in a methanolic iodine solution followed by sputter depth profiling through the oxide towards the interface, but both suffer from their own particular problems. Careful ultramicrotomy followed by STEM used in conjunction with windowless EDX, EELS, or electron diffraction may be more useful where an interphase has developed, but the analysis of the interface on an atomic scale by these methods is still some way off. The most productive approach to probing the interface chemistry directly is the use of thin layers of model compounds deposited from very dilute solutions or even the use of dilute solutions of multi-component commercial products. As an example of the former method, Figure 5.44 shows a set of C 1s spectra recorded from thin films (<2 nm) of poly(methy methacrylate) applied to various oxidized metal substrates.

Subtle differences in the C 1s spectra are observed which are ascribed to the nature of the interactions between the polymer and the oxide substrates. The three substrates are silicon, aluminium, and nickel whose oxides are acidic, weakly basic, and strongly basic respectively. The interactions that occur are shown in Figure 5.45 and are hydrogen bonding (silicon) bidentate interaction (aluminium) and acyl nucleophilic attack (nickel), which are established on the basis of the C 1s spectra of Figure 5.44.

In a similar vein Figure 5.46 shows the N 1s spectrum of diethanolamine (DEA which is a convenient analogue for a cured epoxy resin) adsorbed on oxidised aluminium treated with the adhesion promoter glycidoxypropyl trimethoxy silane (GPS). The two components represent nitrogen with a partial charge, (δ^+), at the lower binding energy and quaternary nitrogen $(-C-NH_2^+-C-)$ represented by the higher binding energy component. This indicates the two different modes of interaction experienced by the DEA molecule. The partial charge results from intermolecular (hydrogen bonding) between adjacent DEA molecules whilst the quaternary component represents a formal interaction between the DEA molecule and the GPS treated aluminium substrate. The change in relative intensities of the higher binding-energy component results from a change in conformation of the adsorbed molecules:

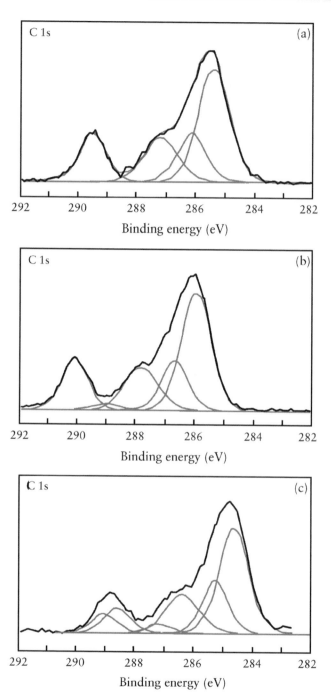

C 1s   (a)

292   290   288   286   284   282
                        Binding energy (eV)

C 1s   (b)

292   290   288   286   284   282
                        Binding energy (eV)

C 1s   (c)

292   290   288   286   284   282
                        Binding energy (eV)

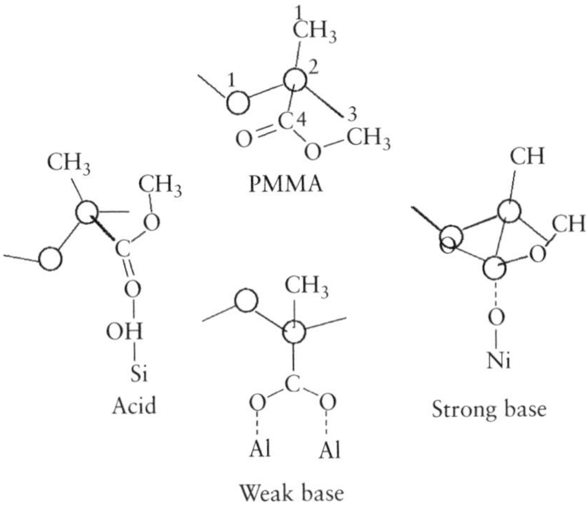

Figure 5.45 Interactions responsible for the fine structure in the C 1s spectra of Figure 5.44

at low concentrations the DEA molecules adopt an orientation in which they lie parallel to the substrate but at higher concentrations a 'bristle-like' orientation is adopted. In the latter case the signal from the interfacial interaction is attenuated, to a certain extent, by the overlying DEA molecule.

In this way a great deal of information has been obtained in recent years concerning the manner of interaction of organic molecules, relevant to adhesives and organic coatings, with oxidized metal substrates. An important complementary technique in this work has been high-resolution time-of-flight secondary ion mass spectrometry (ToF-SIMS). As pointed out in the next chapter, the combination of XPS with ToF-SIMS is very powerful for the comprehensive definition of polymer surface chemistry and the interaction of organic molecules with solid substrates. XPS can also be used to monitor the capacity of a solid surface for species in the liquid phase by the construction of adsorption isotherms from XPS (or ToF-SIMS) data. This has the advantage that the uptake at a variety

Figure 5.44 C 1s spectra of thin films of PMMA applied to oxidized metal substrates of different acid base character: (a) silicon, acidic, (b) aluminium, weakly basic, (c) nickel, strongly basic (Reprinted from J. F. Watts, in *Handbook of surface and interface analysis* (eds J. C. Riviere and S. Myra) p. 822, Fig. 18 by courtesy of Marcel Dekker Inc.)

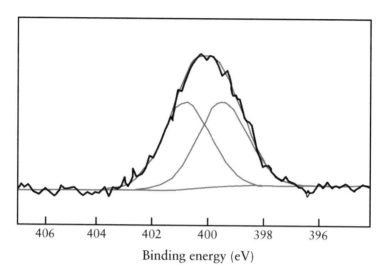

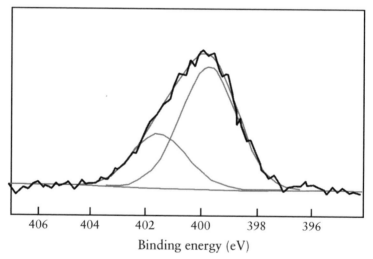

Figure 5.46 N 1s spectrum of diethanolamine adsorbed on oxidized aluminium (reprinted with permission from M.-L. Abel *et al.* Langmuir, **16**, 6510–6518, Fig. 5. Copyright 2000 American Chemical Society)

of solution concentrations can be recorded directly by measurement of the surface composition – the uptake curve is simply the surface composition as a function of the solution concentration.

The analysis of a failed interface is routinely carried out by electron spectroscopy and the definition of adhesive or cohesive failure, on a

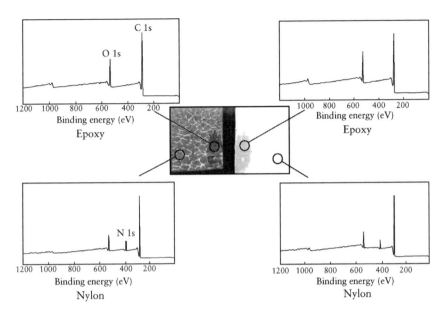

Figure 5.47 Failure of a TFS/epoxy-phenolic/nylon system

molecular scale, has become a straightforward matter for those working in the field. The spectra of Figure 5.47 represent analyses, at a spot size of 400 μm, taken from the failure illustrated in the same figure. The system concerned is a tin-free steel substrate coated with an epoxy-phenolic lacquer which in turn has been joined, by a hot-melt process, to a nylon substrate. The practical application of this system is the bonding of gas capsules ('Widgets') to the internal surfaces of metal beverage cans. The visual appearance of the complementary mirror images of the failure surface is quite striking. The nylon side of the failure is seen as a white colouration (right-hand side of Figure 5.47) with a dark patch on the lower left; the coated TFS substrate (left-hand side of Figure 5.47) is darker with a white lace-like deposit together with a much darker region on the lower right. The spectra recorded from the minority regions are characteristic of epoxy and represent cohesive failure within the epoxy lacquer. The failure between nylon (white) and epoxy is, however, shown to be cohesive within the thermoplastic, as shown by the similarity of the lower spectra of Figure 5.47. Visual observations alone would have identified the failures as interfacial between the lacquer and TFS, and interfacial between the nylon and the lacquer.

As indicated by the examples chosen above it is XPS rather than AES which is most widely used in adhesion studies, often in conjunction with ToF-SIMS. Another area of intense activity over the last decade or so has been their use in studies of composite materials. XPS has been used for some time to assess the surface acidity of carbon fibres and the level of sizing, *in situ* fracture within the spectrometer can now be achieved quite readily on carbon fibre composite materials. Already, both XPS and AES have been used to study the interfacial region in metal matrix composites and it appears that, in some cases, minor elements from the alloy matrix may segregate to the (ceramic) fibre surface.

While the examples cited in this chapter have dealt exclusively with the techniques described earlier in this book, it would be a very narrow-minded scientist who did not make full use of the plethora of advanced analytical techniques which are becoming available. In the next chapter the more common ones are described and comparisons drawn with XPS and AES.

6 Comparison of XPS and AES with Other Analytical Techniques

A compilation some years ago of physical examination and analytical techniques identified over 150 methods which could be used for materials analysis. The set of acronyms assigned to these methods is now vast and, inevitably, confusion has arisen for the surface scientist. For example, SAM stands for scanning Auger microscopy but an equally acceptable meaning is scanning acoustic microscopy. The majority of these techniques are specialist methods requiring careful specimen preparation and experimentation, others are applicable to a fairly limited portion of the periodic table or accept specimens in only one particular form. The aim of this section is to compare the subjects of this text with other analytical methods which are available within research institutes and academia. We shall exclude those processes which yield structural information, such as X-ray and electron diffraction; also excluded are the various vibrational spectroscopies which yield essentially molecular rather than elemental analyses, such as infra-red and Raman spectroscopy. The classification of analysis methods may be carried out in several ways but, for the time being, we shall consider them in terms of primary (incident) and secondary (emitted) radiations; Table 6.1 lists ten, of the many possible methods, which we shall consider.

The acronyms used have the following meanings:

EDX   energy dispersive X-ray analysis

EELS   electron energy-loss spectroscopy

ISS   ion scattering spectroscopy

Table 6.1   Features of various analytical methods discussed in the text

	Incident radiation	Emitted radiation	Property monitored	Elements detectable	Depth of analysis	Spatial resolution	Information level E = elemental C = chemical	Quantification*	Applicability to inorganics*	Applicability to organics*
AES	e⁻	e⁻	Energy	Li on	3–10 nm	<12 nm	E (C)	✓	0	X
EDX	e⁻	X-ray	Energy	Be on	1 µm	1 µm	E	✓	0†	X†
EELS	e⁻	e⁻	Energy	Li on	Depends on foil thickness	10 nm	E	0	✓	X
ISS	ions	ions	Energy	Li on	Outer atom layer	100 µm	E	0	✓	0
LAMMS	laser	ions	Mass	All	0.5 µm	1 µm	E, C	X	✓	✓
RBS	ions	ions	Energy	Li on	1 µm	1 mm	E	X	✓	✓†
SIMS (static)	ions	ions	Mass	All	1.5 nm	1 µm	C (E)	X	✓	✓
SIMS (dynamic)	ions	ions	Mass	All	See text	50 µm	E	0	✓	X
SIMS (imaging)	ions	ions	Mass	All	See text	50 nm	C (E)	X	0	0
XPS	X-rays	e⁻	Energy	He on	3–10 nm	STD 1 mm² small area: 10 µm imaging XPS: <3 µm	E, C	✓	✓	✓

*✓ = very good, 0 = reasonable, X = poor
†Without conductive coating
‡Cryo-stage required

LAMMS   laser ablation microprobe mass spectrometry

RBS   Rutherford backscattering spectrometry

STEM   scanning transmission electron microscopy

SIMS   secondary ion mass spectrometry

TEM   transmission electron microscopy

6.1 X-ray Analysis in the Electron Microscope

The addition of an X-ray analyser (either energy or wavelength disper-sive, EDX or WDX) to a scanning electron microscope provides an electron probe microanalysis facility widely used in all branches of re-search and development. This provides a very flexible means of micro-analysis and, in the conventional EDX mode, elements from beryllium onwards can be detected; with a WDX spectrometer, resolution and sensitivity are improved.

Recent advances in EDX detectors have extended the range to much lighter elements, but the vacuum requirements are more stringent to prevent icing up of the detector. However, although it has now become possible to undertake light-element analysis by the consideration of emitted characteristic X-rays, such analyses are essentially probing bulk composition. The 'interaction volume' of the electron beam with the sample determines both the lateral resolution (in the analytical mode) and the depth of analysis; this is a function of primary beam energy but will invariably be of the order of a micrometer. By reducing beam en-ergy, the depth of analysis may be reduced to as little as 500 nm but there is a limit to this approach; although an electron beam of 1 keV will only have a small penetration depth it will not excite X-rays of analyt-ical use in the conventional sense. The characteristic electron inelastic mean free path for electrons of analytical use in electron spectroscopy is compared with that for the higher energies used as primary radiation in the electron microscope in Figure 6.1. Thus the analysis depth in X-ray analysis is determined by the energy of the primary radiation (the elec-tron beam) whereas in electron spectroscopy it is the energy of the secondary radiation (emitted electrons) which controls this parameter.

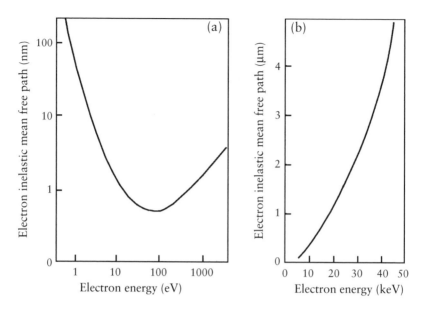

Figure 6.1 Electron mean free paths of energies used in electron spectroscopy (a), and of the primary beam energies used in electron microscopy (b)

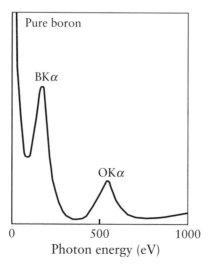

Figure 6.2   Light-element analysis (the sample is oxidized boron) in the TEM by EDX

In the TEM, the use of a thin foil specimen immediately defines the depth of analysis, as the pear-shaped interaction volume is abruptly truncated. This dramatically improves spatial resolution, although there will be some degradation of spatial resolution as a result of electron beam interaction with the sample. An EDX spectrum of pure boron

Figure 6.3 A modern Auger electron spectrometer fitted with an energy dispersive X-ray detector

(with a thin oxide film present), obtained in a TEM, is shown in Figure 6.2; the BKα line is very clear in the spectrum.

In the scanning Auger microscope the analysis depth is determined by the energy of the outgoing electrons as described in Chapter 1 and the spatial resolution depends on the size of the electron probe. Thus, the spatial resolution attainable with SAM can be an order of magnitude better than that recorded with SEM/EDX. With the development of sub-micron features in microelectronics SAM is finding a new use: as a high-resolution chemical imaging facility, the emphasis no longer being on the need for a *surface* analysis.

Although X-rays do show a small chemical shift with oxidation state, this feature is not employed in analytical X-ray analysis of the type used in electron microscopes; thus EDX only provides elemental information unlike the additional chemical information provided by XPS and AES.

The addition of an EDX facility to a surface analysis spectrometer is worthy of consideration. In the SAM mode it is possible to acquire both Auger (surface) and X-ray (bulk) chemical maps of the specimen. A typical set-up is shown in Figure 6.3.

In conjunction with XPS, an X-ray detector is able to provide good quality fluorescence spectra (XRF) to within about 2 keV of the X-ray source operating potential. As this will usually be around 15 keV, X-ray spectra up to 13 keV can be obtained. This method is particularly useful for insulating specimens not amenable to analysis by AES/EDX such as paint films or some catalysts. XRF is also useful in XPS depth profiling where a global analysis of specimen chemistry can be achieved before segregation or interfacial effects are studied in detail.

6.2 Electron Analysis in the Electron Microscope

As well as the possibility of X-ray analysis in the TEM and STEM, it is possible to analyse the energy of the transmitted beam which is the basis of electron energy-loss spectroscopy (EELS). As an electron beam passes through an electron transparent specimen, it is able to eject electrons whose binding energies are less than that of the primary beam energy. By recording the depletion of the primary beam energy with an electron spectrometer positioned below the specimen, an energy-loss spectrum

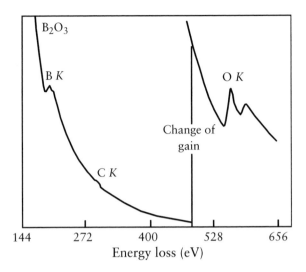

Figure 6.4 EELS analysis of B_2O_3 in the TEM

can be obtained. Such a spectrum will have characteristic edges at energy-loss values equivalent to the core electron's binding energy. As any electron of energy loss greater than the binding energy can cause ionization, the edge has an abrupt start at the binding energy but tails off as the energy loss increases and could decay until the primary beam energy is reached (i.e., almost total loss of energy). In practice, the cross-section falls as E^{-4} and the resultant inner shell edge resembles a triangle, as can be seen in the EELS spectrum of B_2O_3 in Figure 6.4.

The energies of these characteristic edges observed in EELS are close to the binding energies used in XPS; e.g., carbon K edge is 284 eV, magnesium K edge is 1305 eV, copper $L_{2,3}$ edge (convolution of Cu $2p_{3/2}$ and Cu $2p_{1/2}$) is 941 eV. Modern EELS spectrometers collect data in a parallel manner and the technique is referred to as parallel EELS (PEELS). Using PEELS it is possible, in principle, to deduce chemical-state information in a manner similar to XPS and this aspect of the technique is illustrated in Figure 6.5 which shows the Mn $L_{2,3}$ edge for a series of manganese compounds.

The major problems of quantifying EELS spectra are accurately stripping off the background and deconvoluting overlapping or adjacent peaks. These difficulties have now been largely overcome and the outlook appears very promising. In addition, PEELS mapping can be carried out in the STEM, producing chemical images of very light elements.

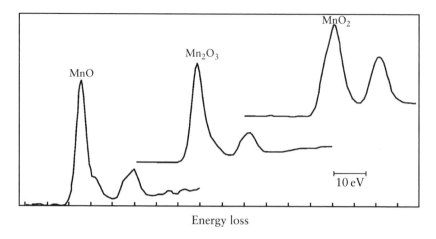

Energy loss

Figure 6.5 PEELS data for a series of manganese compounds (data courtesy of Dr. Vlad Stolojan, University of Surrey, UK)

Once again, imaging EELS does not generally have the chemical specificity of emitted electron spectroscopy, although some elements do show a chemical shift analogous to that of XPS as indicated in Figure 6.5. Specimen preparation can be complex and/or tedious. The one great advantage of EELS (shared by EDX or WEDX in the STEM) is that of spatial resolution which is determined by the resolution of the microscope plus the 'beam spreading' contribution. The depth of analysis is clearly defined by the foil thickness, which will be of the order of tens of nanometers. The very important role that EELS has to play is in the identification of very fine features containing light elements, such as oxide, carbide, and nitride precipitates in steels.

6.3 Mass Spectrometry for Surface Analysis

In parallel with the growth and development of XPS and AES over the last three decades, the technique of secondary ion mass spectrometry (SIMS) has also been evolving. The basic principle of SIMS is the bombardment of a surface with a beam of energetic ions and the subsequent mass analysis of the sputtered ions and cluster ions.

The strengths of this method are:

- all elements including hydrogen can be detected,

- as the analysis is based on the mass separation of the secondary particles, it has both isotopic and molecular specificity,

- it has much higher elemental sensitivity than either XPS or AES.

The major disadvantage of SIMS is that quantification of the data is much more difficult than it is with the electron spectroscopies.

There are three modes of operation in SIMS.

1. Static SIMS (SSIMS) is where the low ion flux ($<10^{12}$ ions cm^{-2} per analysis) ensures that the original surface is insignificantly damaged during the analysis. SSIMS can be applied to polymers and other insulating specimens with a great deal of success. A typical static SIMS spectrum, in the range m/z 0–200 for both positive and negative ions of the polymer poly(ethylene terephthalate), is shown in Figure 6.6.

2. Dynamic SIMS (DSIMS) is when sputtering proceeds at a high rate (as in AES/XPS compositional depth profiling) and only those mass fragments of interest are monitored and plotted as a function of sputter time. An example of a DSIMS depth profile is shown in Figure 6.7.

3. Imaging SIMS, which can be achieved in two different ways, relies on either the resolution of the ion beam itself or the ability of the spectrometer to retain spatial information as the secondary ions pass through a magnetic analyser. In the former approach a fine, sub-micron beam, ions, (e.g. gallium), is rastered across the surface and a particular fragment is monitored and used to build up an elemental map of the surface in a sequential manner, as in scanning Auger. This mode of acquisition is referred to as the scanning ion microprobe. The alternative method is known as ion microscopy, in which the entire field of view is imaged simultaneously in the chosen ion species, analogous to parallel imaging in XPS. In this type of imaging SIMS, the lateral resolution attainable is defined by the size of the apertures in the ion optics and, at the ultimate resolution (0.5 μm), by the optical aberrations of the spectrometer itself.

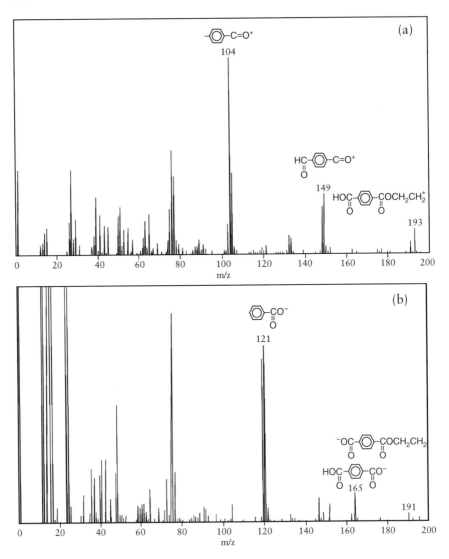

Figure 6.6   SSIMS of poly(ethylene terephthalate): (a) the positive ion spectrum, (b) the negative ion spectrum (the structures of the characteristic cations at m/z = 104, 149, 193 and anions at m/z = 121, 165, 191 are also shown)

All SIMS instruments consist of a vacuum system fitted with at least one ion gun and a mass spectrometer. To optimize performance in the analysis of a wide range of materials a selection of ion sources is required. To maximize the sensitivity of the instrument to positive ions it is customary to use oxygen ions. Similarly, caesium ions will improve

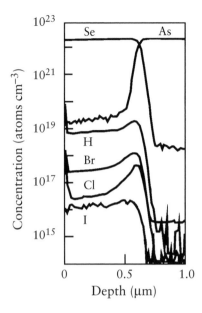

Figure 6.7 SIMS depth profile showing the concentration of impurity elements in an epitaxially grown ZnSe layer on GaAs; the logarithmic concentration scale shows the dynamic range and low detection limits available from a dynamic SIMS measurement (note also that, unlike XPS and AES, SIMS has good sensitivity to hydrogen – data courtesy of MATS UK, Innovation House, Daten Park, Birchwood, Warrington, WA3 6UT, UK)

the sensitivity of the technique to negative ion species. Using caesium can also help with quantification because the probability of forming the ion MCs^+ is less dependent upon the sample than the probability of forming either the M^+ ion or the M^- ion. Liquid metal ion guns are also used in cases where the best spatial resolution is needed. Argon or xenon beams are used when it is necessary to minimize the chemical interaction between the ion beam and the sample.

Three types of mass spectrometer find a use in SIMS.

1. *The magnetic sector* instrument has very high sensitivity and can be operated with high mass resolution; it is the instrument of choice for many dynamic SIMS applications.

2. *The time-of-flight (TOF)* mass spectrometer is ideal for SSIMS applications. Using this instrument, the ion beam is operated in a

pulsed mode and, following every pulse, the complete SIMS spectrum is collected. This minimizes the number of incident ions and therefore maximizes the information available while remaining within the so-called static SIMS regime. The other aspects of TOF spectrometers, which make them ideal for SSIMS, are their very large (in principle, infinite) mass range and their high mass resolution. This type of spectrometer is now becoming important for the dynamic SIMS analysis of ultra-thin films or semiconductor materials implanted with dopants at very low energy.

3. *The quadrupole mass spectrometer* is compact, versatile and relatively inexpensive. In the past, its versatility led to it being used on stand-alone SIMS instruments because it can produce reasonable data in both static and dynamic SIMS measurements. However, with the advent of TOF SIMS, users of the technique have polarized preferring either magnetic sector instruments for dynamic SIMS or TOF SIMS for static SIMS. The quadrupole instrument is now used only as a secondary technique on other surface analysis instruments or for highly specific applications as a stand-alone instrument for dynamic SIMS.

A SIMS analysis is extremely surface sensitive, having an analysis depth in the static SIMS mode of around 1 nm. With dynamic SIMS, the material is being sputtered at such a rate that the actual surface is constantly being eroded and the term 'depth of analysis' becomes less important; it is now the depth resolution that is the parameter by which the technique is judged.

The detection limit for elemental species is probably one of the best obtainable with the methods discussed in this chapter, being of the order of p.p.b in favourable cases, such as boron implanted into silicon. Quantification can be carried out quite accurately by comparison of the specimen being examined with standards of very close composition but, for routine analysis of unknown specimens, quantification will not usually be attempted.

One method which has been used to improve quantification, albeit with much inferior detection limits, is secondary neutral mass spectroscopy (SNMS). Using this technique, the secondary ions are rejected from the secondary beam and a proportion of the neutrals are ionized and detected using the mass spectrometer. This removes the influence of the sample matrix upon the ionization probability and therefore makes quantification more reliable. The ionization step can be accomplished using electron beam, plasma or laser post ionization.

A method closely associated with SIMS, in that the mass analysis of emitted ionized atoms, molecules, and clusters is undertaken, is laser ablation microprobe mass spectrometry (LAMMS). The lateral and depth resolutions attainable with LAMMS are dependent on both the power and the focus conditions of the pulsed laser beam.

A fully focused beam will provide very good spatial resolution but as the power density at the specimen surface is very high the crater formed will be relatively deep; this is the so-called 'hard' laser ionization mode of LAMMS. For the same power output a fully defocused beam will create a crater which has a larger diameter but is much shallower; this is the 'soft' laser desorption mode (0.1 μm is probably the best 'sampling depth' achievable with LAMMS at present). It is interesting to note that this technique was originally developed as a bulk microprobe instrument complementary to EDX and giving elemental and isotopic information down to hydrogen. It is available in both the reflection mode, analogous to SEM/EDX, and transmission configuration, the TEM/EDX equivalent. Although the mass spectrum obtained by LAMMS does not give an *exclusive* surface analysis in the same way that SSIMS, XPS, or AES does, it will often provide valuable information concerning the surface phases. This is because, in the ablation process, the volatilized material of the crater will necessarily include that at the very surface. The high sensitivity of mass analysis techniques such as SIMS and LAMMS ensure that any unsuspected elements present at the surface are clearly defined in the resultant mass spectrum. The major advantages of LAMMS at the present time are its ability to act as a light element/isotopic specific microprobe, and the rapidity with which it can profile thin films; the depth of each laser shot can be matched to film thickness by varying the operating conditions from 'soft' to 'hard'. Unlike SEM and SIMS, insulating samples can be analysed without any special sample preparation or charge compensation being required. The main problem is quantification of the resultant spectrum, although semi-quantitative data can be achieved quite successfully using standards of similar composition.

6.4 Ion Scattering

In this section, we shall briefly consider the scattering of incident ions by a solid sample. This forms the basis of two, markedly different, analytical techniques. In the case of low-energy ions (0.2–3 keV) the method is known as ion scattering spectroscopy (ISS) or, more correctly, as low-energy ion scattering spectroscopy (LEISS). LEISS analysis is extremely surface sensitive. When higher energies (1–5 MeV) are used the primary ions are backscattered from the target atoms and the resultant spectrum can be interpreted not only in terms of elements present but also as a depth profile to a depth of about 1 µm. High-energy ion scattering spectrometry (HEIS) is generally referred to as Rutherford backscattering spectrometry (RBS).

LEISS is often performed on a surface analysis system equipped for electron spectroscopy, the electron energy analyser being capable of operation in either negative detection mode (for XPS or AES) or positive detection mode for scattered ions. The specimen is bombarded by a monoenergetic ion beam (usually He^+, Ne^+, or Ar^+), the ions are scattered elastically from the atoms of the outermost layer of the solid and their energy is measured by the energy analyser. The process is illustrated in Figure 6.8 in which the noble gas ion has a mass M_1 and energy E_1 before colliding with an atom of mass M_2 at the surface of the sample. The collision causes the trajectory of the ion to change by an angle θ (the scattering angle) and its energy to change to E_2.

By applying the conservation laws, it can be shown that:

$$\frac{E_2}{E_1} = \left[\frac{\cos\theta \pm (q^2 - \sin^2\theta)^{1/2}}{1 + q} \right]^2$$

where $q = M_2/M_1$. In the special case of a scattering angle of 90°, the equation simplifies to:

$$\frac{E_2}{E_1} = \frac{M_2 - M_1}{M_2 + M_1}$$

The ion scattered spectrum takes the form of the intensity of the scattered primary beam as a function of its energy normalized to the primary beam energy (E_2/E_1), as shown in Figure 6.9. One of the main

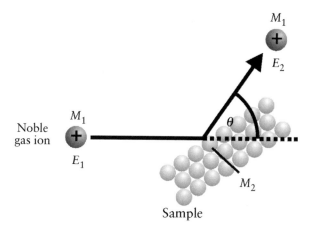

Figure 6.8 The ion scattering process

difficulties of employing ISS stems from the very surface sensitive nature of the analysis and surface contamination may become a serious problem. There are various ways of circumventing this involving *in situ* sample preparation by heating, ion sputtering, or cleaving.

The choice of primary ion is important in LEISS because, as can be seen from the above expression, only atoms whose mass is higher than that of the primary ion can be detected. Therefore helium ions will, in principle, provide detection of the greatest mass range. If good mass resolution is required, the mass of the primary ion should be as close as possible to the mass of the surface atom. It is often helpful to analyse the same sample using two different primary ion species. Figure 6.9 shows two LEISS spectra from a phosphor bronze sample. Using He^+ ions, the copper, tin, and oxygen impurity atoms can be detected. When Ar^+ ions are used instead of He^+, the two isotopes of copper can be separated.

In the case of Rutherford backscattering (RBS), the energy of the primary ion beam is much higher and the experimental set-up will include some form of accelerator to provide ions of sufficient energy (around 2 MeV). The RBS experiment consists of bombarding the surface of the specimen with these high-energy ions (usually He^{++}) and simultaneously measuring the energy of the backscattered ion. The ions have a very small diameter, because they consist only of a nucleus, and are able to penetrate the lattice of the sample. A high proportion of the ions are implanted into the solid but some are scattered and detected.

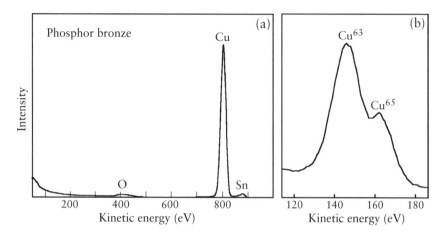

Figure 6.9   ISS spectrum from a phosphor bronze surface: (a) using helium ions and showing the wide mass range, and (b) using argon ions and showing the two copper isotopes

Although the energy of the ions is very much higher in RBS than it is in LEISS, the scattering equations are very similar.

The primary ions that interact with light elements become backscattered ions of low energy, and those primary ions that interact with higher masses are scattered with a higher energy. However, scattering takes place not only at the surface, as in the case of LEISS, but many atomic layers into the sample and ions undergo attenuation on their way both into and out of the specimen. Consequently, the RBS spectrum contains not only information concerning the identity of the atoms within the interaction volume but also their position in relation to the free surface, allowing a depth profile to be constructed from the RBS spectrum. Because of this convolution of elemental and depth information in the final spectrum, care must be taken in interpretation, and it is usual to restrict its application to systems with only a few known elements present. A simulation programme will then be employed to reconstruct the depth distribution of the elements in the specimen accurately. RBS is widely used in the analysis of dopants and thin films used in the microelectronics industry. Quantification can be performed accurately from first principle methods using experimental parameters and scattering cross-sections; detection limits are typically in the 100 p.p.m. range, but can vary very widely depending on the matrix material. The

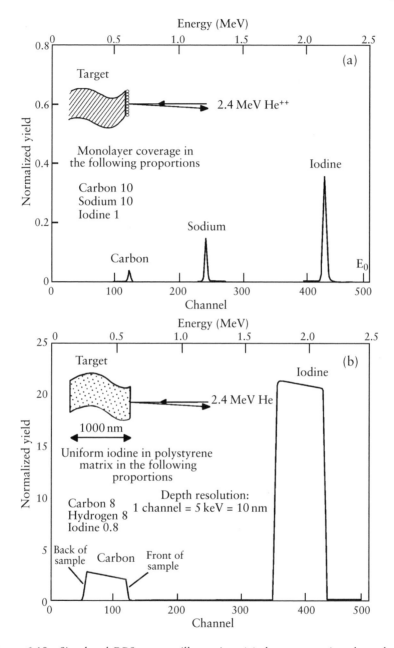

Figure 6.10 Simulated RBS spectra illustrating: (a) the cross-section dependence on atomic number (varies with Z^2), and (b) depth profiling capabilities through a 1 mm film

calculated RBS spectrum of Figure 6.10(a) illustrates the variation in cross-section with increasing mass for a monolayer of material (carbon 10: sodium 10: iodine 1) deposited on a substrate which does not contribute to the spectrum. The depth dependency of the RBS spectrum is seen in Figure 6.10(b), where the specimen is a 1 μm thick deposit of material (carbon 8: hydrogen 8: iodine 0.8). In this case, the individual peaks have broadened considerably and the front and back of the sample can be identified, thus providing a non-destructive depth profile of the sample on a micron scale.

6.5 Concluding Remarks

Of the methods discussed in this chapter, those based on conventional electron microscopes, especially EDX, are readily available in universities and research establishments and most scientists with a need for microanalysis will have encountered them. The other methods are less widely available and general access to them, outside of corporate research laboratories and universities, can only be obtained at one of the specialist surface analysis consultancy services which are available worldwide.

Glossary

Abbreviations

AES	Auger electron spectroscopy
ARXPS	angle resolved XPS
CMA	cylindrical mirror analyser
eV	electron volts
EELS	electron energy-loss spectroscopy
ESCA	electron spectroscopy for chemical analysis
FABMS	fast-atom bombardment mass spectrometry
FWHM	full-width at half maximum
HSA	hemispherical sector analyser
LEISS	low-energy ion scattering spectrometry
RBS	Rutherford backscattering spectrometry
SAM	scanning Auger microscope
SEM	scanning electron microscope
SIMS	secondary ion mass spectrometry
SNMS	sputtered neutral mass spectrometry
SSA	spherical sector analyser
TOF or ToF	time of flight
UPS	ultraviolet photoelectron spectroscopy
XPS	X-ray photoelectron spectroscopy

Surface Analysis Methods[6]

Auger Electron Spectroscopy
A method in which an electron spectrometer is used to measure the energy distribution of Auger electrons emitted from a surface.

Electron Spectroscopy for Chemical Analysis (ESCA)
A method encompassing both AES and XPS. The term ESCA is falling out of use as, in practice, it was only used to describe situations more clearly defined by the term X-ray photoelectron spectroscopy (XPS). The latter term has been preferred.

Secondary Ion Mass Spectrometry (SIMS)
A method in which a mass spectrometer is used to measure the mass-to-charge quotient and abundance of secondary ions emitted from a sample as a result of bombardment by energetic ions

Ultraviolet Photoelectron Spectroscopy (UPS)
A method in which an electron spectrometer is used to measure the energy distribution of photoelectrons emitted from a surface irradiated by ultraviolet photons. Ultraviolet sources in common use include various types of discharges that can generate the resonance lines of various gases (e.g., the He I and He II emission lines at energies of 21.2 eV and 40.8 eV, respectively). For variable energies, synchrotron radiation is used.

X-ray Photoelectron Spectroscopy (XPS)
A method in which an electron spectrometer is used to measure the energy distribution of photoelectrons and Auger electrons emitted from a surface irradiated by X-ray photons. X-ray sources in common use are Al and Mg unmonochromated Kα X-rays at 1486.6 and 1253.6 eV, respectively. Modern instruments also use monochromated AlKα X-rays. Some instruments make use of various X-ray sources with other anodes or of synchrotron radiation.

---

[6]The terms and definitions taken from ISO 18115 are reproduced with the permission of the International Organization for Standardization, ISO. This standard can be obtained from any ISO member and from the website of the ISO Central Secretariat at the following address: http://www.iso.ch/

Terms Used in Surface Analysis[6]

Adventitious Carbon Referencing
Determining the charging potential of a particular sample from a comparison of the experimentally determined C 1s binding energy, arising from adsorbed hydrocarbons on the sample, with a standard binding energy value. A nominal value of 285.0 eV is often used for the binding energy of the relevant C 1s peak, although some analysts prefer specific values in the range 284.6 eV to 285.2 eV depending on the nature of the substrate.

Angle Lapping
Sample preparation in which a sample is mechanically polished at an angle to the original surface. This angle may often be less than 1° so that depth information with respect to the original surface is transformed to lateral information.

Angle of Emission
Angle between the trajectory of a particle or photon as it leaves a surface and the local or average surface normal.

Angle of Incidence
Angle between the incident beam and the local or average surface normal.

Angle resolved XPS (ARXPS) or angle dependent XPS
A procedure in which X-ray photoelectron intensities are measured as a function of the angle of emission.

Angle, Take-off
Angle between the trajectory of a particle as it leaves a surface and the local or average surface plane.

Attenuation Length
Quantity λ in the expression $\Delta x/\lambda$ for the fraction of a parallel beam of specified particles or radiation removed in passing through a thin layer Δx of a substance in the limit as Δx approaches zero, where Δx is measured in the direction of the beam. The intensity or number of particles in the beam decays as $\exp(-x/\lambda)$ with the distance x.

Auger Electron
Electron emitted from atoms in the Auger process.

Auger Electron Spectrum
Plot of the Auger electron intensity as a function of the electron kinetic energy, usually as part of the energy distribution of detected electrons.

Auger Parameter
Kinetic energy of a narrow Auger electron peak in a spectrum minus the kinetic energy of the most intense photoelectron peak from the same element. *In this text we use the modified Auger parameter (α), numerically equal to the sum of the kinetic energy of the Auger transition and binding energy of the photoelectron peak i.e.*

$$\alpha = KE_{\text{Auger}} - KE_{PE} + h\nu$$

Auger Process
Relaxation, by electron emission, of an atom with a vacancy in an inner electron shell, the emitted electrons have characteristic energies, defined by the Auger transition.

Auger Transition
Auger process involving designated electron shells or sub-shells. The three shells involved in the Auger process are designated by three letters. The first letter designates the shell containing the initial vacancy and the last two letters designate the shells containing electron vacancies left by the Auger process (for example, KLL, and LMM). When a valence electron is involved the letter V is used (for example, LMV and KVV). When a particular sub-shell involved is known this can also be indicated (for example, KL_1L_2). Coupling terms may also be added, where known, to indicate the final atomic state (for example, $L_3M_{4,5}M_{4,5};{}^1D$).

Background, Inelastic
Intensity distribution in the spectrum for particles originally at one energy but which are emitted at lower energies due to one or more inelastic scattering process.

Back-scattered Electron
Electron originating in the incident beam which is emitted after interaction with the sample.

Ball Cratering

A procedure in which the sample is abraded by a sphere in order to expose compositional changes in layers below the original surface with the intent that the depth of those layers can be related to the lateral position in the crater created by the abrasion.

Beam Current

Quotient of dQ by dt, where dQ is the quantity of electric charge of a specified polarity in the beam passing in the time interval dt

$$I = dQ/dt$$

Beam Current Density

Quotient of dI by dA where dI is the element of beam current incident on an area dA at right angles to the direction of the beam

$$J = dI/dA$$

Beam Diameter

For a particle beam of circular cross-section the full width of the beam at half maximum intensity measured in a plane normal to the beam direction.

Binding Energy

Energy that must be expended in removing an electron from a given electronic level to the Fermi level of a solid or to the vacuum level of a free atom or molecule.

Bremsstrahlung

Photon radiation emitted from a material due to the deceleration of incident electrons within that material. The bremsstrahlung radiation has a continuous spectral distribution up to the energy of the incident electrons. In XPS, the bremsstrahlung from a conventional X-ray source with an Al or Mg anode leads to a continuous photoelectron background. This radiation may also photoionize inner shells which would be energetically impossible by characteristic Al or Mg Kα X-rays. As a result, Auger electron features may appear at negative binding energy values and, in addition, the intensities of other Auger electron features may be greater than if the inner shell vacancies had been created only by the characteristic X-rays. The bremsstrahlung-excited Auger electron features can be helpful for determining the various Auger parameters needed to identify chemical states.

Chemical Shift
Change in peak energy arising from a change in the chemical environment of the atom.

Compositional Depth Profile
Chemical or atomic composition measured as a function of distance normal to the surface.

Constant ΔE Mode (constant analyser energy mode, CAE mode or fixed analyser transmission mode, FAT mode)
Mode of electron energy analyser operation which varies the electron retardation but keeps the pass energy constant in the energy dispersive portion of the analyser. This mode is often used in XPS to maintain a high- and constant-energy resolution throughout the spectrum.

Constant $\Delta E/E$ Mode (constant retardation ratio mode, CRR mode or fixed retardation ratio mode, FRR mode)
Mode of electron energy analyser operation which varies the retarding potential so that the pass energy in the energy dispersive portion of the analyser is a constant fraction of the original *vacuum level* referenced kinetic energy. This mode is often used in AES to improve the signal-to-noise ratio for high-energy emitted electrons at the expense of spectral resolution.

Depth Profiling
Monitoring of signal intensity as a function of a variable which can be related to distance normal to the surface usually, sputtering time.

Depth Resolution
Depth range over which a signal changes by a specified quantity when reconstructing the profile of an ideally sharp interface between two media or a delta layer in one medium. The precise quantity to be used depends on the signal function with depth. However, for routine analytical use, a convention of the depth at an interface over which the signal

from an overlayer or a substrate changes from 16 per cent to 84 per cent of their total variation between plateau values, is often used in AES and XPS.

Detection Limit
Smallest amount of an element or compound that can be measured under specified analysis conditions.

Differential Spectrum
Differential of the direct spectrum with respect to energy, E, by an analogue electrode modulation method or by numerical differentiation of that spectrum.

Direct Spectrum
Intensity of electrons transmitted and detected by a spectrometer with a dispersing energy analyser, as a function of energy E.

Electron Retardation
Method of measuring the kinetic energy distribution by retarding the emitted electrons before or within the electron energy analyser.

Grazing Exit (or Glancing Exit)
Geometrical arrangement in which the angle of the scattered (or emitted) particles is near 90° from the normal to the sample surface. This configuration generally results in improved surface sensitivity and may also improve depth resolution.

Grazing Incidence (or Glancing Incidence)
Geometrical arrangement in which the angle of the incident particles is near 90° from the normal to the sample surface.

Inelastic Mean Free Path, Electron
Average distance that an electron with a given energy travels between successive inelastic collisions.

Inelastic Scattering
Interaction between a moving energetic particle and a second particle or assembly of particles in which the total kinetic energy is not conserved.

Kinetic energy is absorbed in solids by various mechanisms, for example inner shell ionization, plasmon and phonon excitation and bremsstrahlung generation. These excitations usually lead to a small change in direction of the moving particle.

Information Depth
Maximum depth, normal to the surface, from which useful information is obtained.

Interface Width, Observed
Distance over which a 16 per cent to 84 per cent, or 84 per cent to 16 per cent, change in signal intensity is measured at the junction of two dissimilar matrices, the thicknesses of which are more than six times that distance.

Lateral Resolution
Distance measured either in the plane of the sample surface or in a plane at right angles to the axis of the image-forming optics over which changes in composition can be separately established with confidence.

Line Scan
Plot of the output signal intensity from the spectrometer, the signal intensity from another detector, or processed intensity information from the available software along a line corresponding to a line on the sample surface.

Map or Image
Two-dimensional representation of the sample surface where the information at each point in the representation is related to the output signal from the spectrometer, the signal from another detector, or processed intensity information from the available software. By convention, map is usually applied to cases where the information is primarily composition-specific and image to those where it is primarily topographic. Map intensities may be presented in a normalized fashion to have the maximum and minimum signal intensities set at, for example, full white and full black, respectively, or on a colour scale. The contrast scale should be defined.

Modified Auger Parameter
Auger parameter, to which photon energy is added (to avoid a negative quantity). Referred to in this text as Auger parameter, α.

$$\alpha = KE_{Auger} - KE_{PE} + h\nu$$

i.e.

$$\alpha = KE_{Auger} + BE_{PE}$$

Monolayer
Complete coverage of a substrate by one atomic or molecular layer of a species.

Multiplet Splitting
Splitting of an Auger electron line into two or more components caused by the interactions of the atomic vacancies created by the Auger process, *or* splitting of a photoelectron line caused by the interaction of the unpaired electron created by photoemission with other unpaired electrons in the atom.

Noise
Time-varying disturbances superimposed on the analytical signal with fluctuations leading to uncertainty in the signal intensity.

Noise, Statistical
Noise in the spectrum due solely to the statistics of randomly-detected single events. For single particle counting systems exhibiting Poisson statistics, the standard deviation of a large number of measures of an otherwise steady count rate, N, each in the same time interval, is equal to the square root of N.

Pass Energy
Mean kinetic energy of the detected particles in the energy dispersive portion of the energy analyser.

Peak Fitting
A procedure whereby a spectrum, generated by peak synthesis, is adjusted to match a measured spectrum. A least-squares optimization procedure is generally used in a computer program for this purpose.

Peak Synthesis
Procedure whereby a synthetic spectrum is generated using either model or experimental peak shapes in which the number of peaks, the peak shapes, peak widths, peak positions, peak intensities and the background shape and intensity are adjusted for peak fitting.

Peak-to-Background Ratio (or Signal-to-Background Ratio)
Ratio of the maximum height of the peak above the background intensity to the magnitude of that background intensity.

Peak Width
Width of a peak at a defined fraction of the peak height. The most common measure of peak width is the full width of the peak at half maximum (FWHM) intensity.

Photoelectric Effect
Interaction of a photon with bound electrons in atoms, molecules, and solids, resulting in the production of one or more photoelectrons.

Photoelectron X-ray Satellite Peaks
Photoelectron peaks in a spectrum resulting from photoemission induced by characteristic minor X-ray lines associated with the X-ray spectrum of the anode material.

Photoemission
Emission of electrons from atoms or molecules caused by the photoelectric effect.

Plasmon
Excitation of valence band electrons in a solid in which collective oscillations are generated.

Primary Electron
Electron extracted from a source and directed to a sample.

Profile, Depth
Chemical or elemental composition, signal intensity, or processed intensity information from the available software measured in a direction normal to the surface.

Raster
Two-dimensional pattern generated by the deflection of a primary beam. Commonly used rasters cover square or rectangular areas.

Relative Resolution of a Spectrometer
Ratio of the resolution of a spectrometer at a given energy, mass, or wavelength to that energy, mass, or wavelength.

Secondary Electron
Electron, generally of low energy, leaving a surface as a result of an excitation induced by an incident electron, photon, ion, or neutral particle.

Selected Area Aperture
Aperture in the electron or ion optical system restricting the detected signal to a small area of the sample surface.

Shakeup
Multielectron process in which an atom is left in an excited state following a photoionization or Auger electron process, so that the outgoing electron has a characteristic kinetic energy slightly less than that of the parent photoelectron.

Signal-to-Noise Ratio
Ratio of the signal intensity to a measure of the total noise in determining that signal.

Smoothing
Mathematical treatment of data to reduce the apparent noise.

Spectrometer Transmission Function
Quotient of the number of particles transmitted by the analyser by the number of such particles per solid angle and per interval of the dispersing parameter (e.g., energy, mass, or wavelength) available for measurement as a function of the dispersing parameter.

Spin Orbit Splitting
Splitting of p, d, or f levels in an atom arising from coupling of the spin and orbital angular momentum.

Sputter Depth Profile
Compositional depth profile obtained when the surface composition is measured as material is removed by sputtering.

Sputtering
Process in which atoms and ions are ejected from the sample as a result of particle bombardment.

Sputtering, Preferential
Change in the equilibrium surface composition of the sample which may occur when sputtering multicomponent samples.

Sputtering Rate
Quotient of the amount of sample material removed, as a result of particle bombardment, by time.

Sputtering Yield
Ratio of the number of atoms and ions sputtered from a sample to the total number of incident primary particles.

Vacuum Level
Electric potential of the vacuum at a point in space. In electron spectroscopy, the point in space is taken at a sufficiently large distance outside the sample such that electric fields caused by different work functions of different parts of the surface are zero or extremely small.

Vacuum-level Referencing
Method of establishing the kinetic-energy scale in which the zero point corresponds to an electron at rest at the vacuum level.

Valence Band Spectrum
Photoelectron energy distribution arising from excitation of electrons from the valence band of the sample material.

Work Function
Potential difference for electrons between the Fermi level and the maximum potential just outside a specified surface.

X-ray Monochromator
Device used to eliminate photons of energies other than those in a narrow energy or wavelength band.

Bibliography

Chapter 1

Briggs, D. and Seah, M. P. (1990). *Practical surface analysis by Auger and X-ray photoelectron spectroscopy (Second Edition)*. John Wiley and Sons Ltd, Chichester, UK.

Brundle, C. R. and Baker, A. D. *Electron spectroscopy: theory, techniques and applications*. Vol. 1 (1977), Vol. 2 (1978), Vol. 3 (1979), Vol. 4 (1981). Academic Press Inc., New York, USA.

Carlson, T. A. (1976). *Photoelectron and Auger spectroscopy*. Plenum, New York, USA.

Castle, J. E. (1982). Electron spectroscopy methods. In *Analysis of high temperature materials* (ed. O. Van der Biest), pp. 141–188. Applied Science Publishers Ltd, London, UK.

Ferguson, I. F. (1989). *Auger microprobe analysis*. Adam Hilger Ltd, Bristol, UK.

Nefedov, V. I. (1988). *X-ray photoelectron spectroscopy of solid surfaces*. VSP BV, Utrecht, The Netherlands.

Riviere, J. C. (1982). Auger techniques in analytical chemistry: a review. *The Analyst*, **108**, 649–684.

Riviere, J. C. and Myra, S. (1998). *Handbook of surface and interface analysis*. Marcel Dekker Inc, New York, USA.

Seah, M. P. and Dench, W. A. (1979). Quantitative electron spectroscopy of surfaces: a standard data base for electron inelastic mean free paths in solids. *Surf. Interf. Anal.*, **1**, 2–11.

Siegbahn, K. *et al.* (1967). *ESCA: atomic, molecular, and solid state structure studied by means of electron spectroscopy*. Almqvist and Wiksells, Uppsala, Sweden.

Chapter 2

Barrie, A. (1977). Instrumentation for electron spectroscopy. In *Handbook of ultraviolet and X-ray photoelectron spectroscopy* (ed. D. Briggs), pp. 79–119. Heyden and Sons Ltd, London, UK.

Brooker, A. D. and Castle, J. E. (1986). Scanning Auger microscopy: resolution in time, energy and space. *Surf. Interf. Anal.*, **8**, 113–119.

Brundle, C. R. *et al.* (1974). An ultrahigh vacuum electron spectrometer for surface studies. *J. Elec. Spec.*, **3**, 241–261.

Castle, J. E. and West, R. H. (1980). The utility of bremsstrahlung induced Auger peaks. *J. Elec. Spec.*, **18**, 355–358.

Coxon, P. *et al.* (1990). Escascope – a new imaging photoelectron spectrometer. *J. Elec. Spec.*, **52**, 821–838.

Edgell, M. J., Paynter, R. W. and Castle, J. E. (1985). High energy XPS using a monochromated AgLα source: resolution, sensitivity and photoelectric cross sections. *J. Elec. Spec.*, **37**, 241–256.

Koenig, M. F. and Grant, J. T. (1985). Monochromator versus deconvolution for XPS studies using a Kratos ES300 system. *J. Elec. Spec.*, **36**, 213–225.

Seah, M. P., Anthony, M. T. and Dench, W. A. (1983). Characterization of computer differentiation of AES spectra and its relation to differentiation by the modulation technique. *J. Phys. E.*, **16**, 848–857.

Sherwood, P. M. A. (1983). Data analysis in X-ray photoelectron spectroscopy. In *Practical surface analysis by Auger and X-ray photoelectron spectroscopy* (ed. D. Briggs and M. P. Seah), pp. 445–475. John Wiley and Sons Ltd, Chichester, UK.

Wagner, C. D. and Joshi, A. (1988). The Auger parameter, its utility and advantages: a review. *J. Elec. Spec.*, **47**, 283–313.

Chapter 3

Davis, L. E. *et al.* (1978). *Handbook of Auger electron spectroscopy*. Physical Electronics Division, Perkin-Elmer Corporation, Eden Prairie, USA.

Fuggle, J. C. and Martenson, N. (1980). Core-level binding energies in metals. *J. Elec. Spec.*, **21**, 275–281.

Hall, P. M. and Morabito, J. M. (1979). Matrix effects in the quantitative Auger analysis of dilute alloys. *Surf. Sci.*, **83**, 391–405.

Hall, P. M., Morabito, J. M. and Conley, D. K. (1977). Relative sensitivity factors for Auger analysis of binary alloys. *Surf. Sci.*, **62**, 1–20.

Seah, M. P. (1979). Quantative AES: via the energy spectrum of the differential? *Surf. Interf. Anal.*, **1**, 86–90.

Seah, M. P. (1980). The quantitative analysis of surfaces by XPS: a review. *Surf. Interf. Anal.*, **2**, 222–239.

Seah, M. P. (1983). A review of quantitative Auger electron spectroscopy. *Scanning Electron Microscopy*, SEM Inc., Chicago, USA, **2**, 521–536.

Wagner, C. D. *et al.* (1979). *Handbook of X-ray photoelectron spectroscopy.* Physical Electronics Division, Perkin-Elmer Corporation, Eden Prairie, USA.

Wagner, C. D. *et al.* (1981). Empirical sensitivity factors for quantitative analysis by electron spectroscopy for chemical analysis. *Surf. Interf. Anal.*, **3**, 211–225.

Chapter 4

Chang, J. P. *et al.* (2000). Profiling nitrogen in ultrathin silicon oxynitrides with angle resolved X-ray photoelectron spectroscopy. *J. Appl. Phys.*, **87**, 4449–4455.

Cumpson, P. J. (1995). Angle resolved XPS and AES: depth resolution limits and a general comparison of depth profile reconstruction methods. *J. Elec. Spec.*, **73**, 25–52.

Hofmann, S. (1983). Depth profiling. In *Practical surface analysis by Auger and X-ray photoelectron spectroscopy* (ed. D. Briggs and M. P. Seah), pp. 141–179. John Wiley and Sons Ltd, Chichester, UK.

Kelly, R. (1985). On the role of gibbsian segregation in causing preferential sputtering. *Surf. Interf. Anal.*, **7**, 1–7.

Lea, C. and Seah, M. P. (1981). Optimized depth resolution in ion-sputtered and lapped compositional profiles with Auger electron spectroscopy. *Thin Solid Films*, **75**, 67–86.

Paynter, R. W. (1981). Modification of the Beer-Lambert equation for application to concentration gradients. *Surf. Interf. Anal.*, **3**, 186–187.

Seah, M. P. (1981). Pure element sputtering yields using 500–1000 eV argon ions. *Thin Solid Films*, **81**, 279–287.

Seah, M. P. and Hunt, C. P. (1983). The depth dependence of the depth resolution in composition-depth profiling with Auger electron spectroscopy. *Surf. Interf. Anal.*, **5**, 33–37.

Smith, G. C. and Livsey, A. K. (1992). Maximum entropy: A new approach to non-destructive depth from angle dependent XPS. *Surf. Interf. Anal.*, **19**, 175.

Walls, J. M., Hall, D. D. and Sykes, D. E. (1979). Compositional-depth profiling and interface analysis of surface coatings using ball-cratering and the scanning Auger microprobe. *Surf. Interf. Anal.*, **1**, 204–210.

Yih, R. S. and Ratner, B. D. (1987). A comparison of two angular dependant ESCA algorithms useful for constructing depth profiles for surfaces. *J. Elec. Spec.*, **43**, 61–82.

Chapter 5

Baer, D. R. (1984). Solving corrosion problems with surface analysis. *Appl. Surf. Sci.*, **19**, 382–396.

Beamson, G. and Briggs, D. (1992). *High resolution XPS of organic polymers: the Scienta ESCA300 Database*. John Wiley and Sons Ltd, Chichester, UK.

Bhasin, M. M. (1975). Auger spectroscopic analysis of the poisoning of commercial palladium-alumina hydrogenation catalyst. *J. Catalysis*, **38**, 218–222.

Briggs, D. (1998). *Surface analysis of polymers by XPS and static SIMS*. Cambridge University Press, Cambridge, UK.

Brinen, J. S. *et al.* (1984). Characterization of fresh and spent HDS catalysts by Auger and X-ray photoelectron spectroscopies. *Surf. Interf. Anal.*, **6**, 68–73.

Castle, J. E. (1986). The role of electron spectroscopy in corrosion science. *Surf. Interf. Anal.*, **9**, 345–356.

Castle, J. E. and Watts, J. F. (1988). The study of interfaces in composite materials by surface analytical techniques. In *Interfaces in polymer, ceramic, and metal matrix composites* (ed. H. Ishida), pp. 57–71. Elsevier Science Publishing Co. Inc., New York, USA.

Diplas, S. *et al.* (2001). XPS studies of Ti-Al and Ti-Al-V alloys using $CrK\beta$ radiation. *Surf. Interf. Anal.*, **31**, 734–744.

Heckingbottom, R. (1986). Perspectives in surface and interface analysis for electronic devices and circuits. *Surf. Interf. Anal.*, **9**, 265–274.

Holloway, P. H. and McGuire, G. E. (1980). Characterization of electronic devices and materials by surface sensitive analytical techniques. *Appl. Surf. Sci.*, **4**, 410–444.

O'Hare, L.-A. *et al.* (2002). Surface physico-chemistry of corona-discharge-treated PET film. *Surf. Interf. Anal.*, **33**, 617–625.

Paynter, R. W. and Ratner, B. D. (1985). The study of interfacial proteins and bimolecules by X-ray photoelectron spectroscopy. In *Surface and interfacial aspects of biomedical polymers* (ed. J. D. Andrade), pp. 189–216. Plenum Press, New York, USA.

Perruchot, C. *et al.* (2002). High resolution XPS of crosslinking and segregation phenomena in hexamethoxymethyl melamine polyester resins. *Surf. Interf. Anal.*, **34**, 570–574.

Pijpers, A. P. and Meier, R. J. (1987). Oxygen-induced secondary substituent effects in polymer XPS spectra. *J. Elec. Spec.*, **43**, 131–137.

Prickett, A. C., Smith, P. A. and Watts, J. F. (2001). ToF-SIMS studies of carbon fibre fracture surfaces and the development of controlled Mode in situ fracture. *Surf. Interf. Anal.*, **31**, 11–17.

Reilley, C. N., Everhart, D. S. and Ho, F. F.-L. (1982). ESCA analysis of functional groups on modified polymer surfaces. In *Applied electron spectroscopy for chemical analysis* (ed. H. Windawi and F. F.-L. Ho), pp. 105–133. John Wiley and Sons, New York, USA.

Riviere, J. C. and Myra, S. (1998). *Handbook of surface and interface analysis.* Marcel Dekker Inc, New York, USA.

Seah, M. P. and Hondros, E. D. (1977). Segregation to interfaces. *Int. Metal. Rev.*, **22**, 262–301.

Watts, J. F. (1985). Analysis of ceramic materials by electron spectroscopy. *J. Microscopy*, **140**, 243–260.

Watts, J. F. (1987). The use of X-ray photoelectron spectroscopy for the analysis of organic coating systems. In *Surface coatings I* (ed. A. D. Wilson, J. W. Nicholson and H. J. Prosser), pp. 137–187. Elsvier Applied Science Publishers Ltd, London, UK.

Watts, J. F. (1988). The application of surface analysis to studies of the environmental degradation of polymer-to-metal adhesion. *Surf. Interf. Anal.*, **12**, 497–503.

Watts, J. F. (1998). Adhesion science and technology. In *Handbook of surface and interface analysis* (ed. J. C. Riviere and S. Myra), pp. 781–734. Marcel Dekker Inc, New York, USA.

Watts, J. F. *et al.* (2001). Segregation and crosslinking in urea formaldehyde resins: A study by high resolution XPS. *J. Elec. Spec.*, **121**, 233–247.

West, R. H. and Castle, J. E. (1982). The correlation of the Auger parameter with refractive index: and XPS study of silicates using ZrLα radiation. *Surf. Interf. Anal.*, **4**, 68–75.

The Proceedings of the Biennial European Conference on Applications of Surface and Interface Analysis (ECASIA) are published as a single bound volume of Surface and Interface Analysis, (ECASIA'99 Vol. 30, ECASIA'01 Vol. 34, earlier conferences are also recorded in this manner). These Proceedings provide a timely overview of the application of surface analysis in all aspects of materials science.

Chapter 6

Baun, W. L. (1981). Ion scattering spectrometry: a versatile technique for a variety of materials. *Surf. Interf. Anal.*, **3**, 243–250.

Budd, P. M. and Goodhew, P. J. (1988). *Light-element analysis in the transmission electron microscope: WEDX and EELS.* Oxford University Press, Oxford, UK.

Castle, J. E. and Castle, M. D. (1983). Simultaneous XRF and XPS analysis. *Surf. Interf. Anal.*, **5**, 193–198.

Chu, W. K., Mayer, J. W. and Nicolet, M.-A. (1978). *Backscattering spectrometry.* Academic Press Inc., New York, USA.

Clarke, N. S., Ruckman, J. C. and Davey, A. R. (1986). The application of laser ionization mass spectrometry to the study of thin films and near-surface layers. *Surf. Interf. Anal.*, **9**, 31–40.

Degreve, F., Thorne, N. A. and Lang, J. M. (1988). Metallurgical applications of SIMS. *J. Mater. Sci.*, **23**, 4181–4208.

Goodhew, P. J. and Castle, J. E. (1983). A survey of physical examination and analysis techniques. *Inst. Phys. Conf. Ser. No. 68 (EMAG)*, pp. 515–522.

Goodhew, P. J. and Humphreys, F. J. (1988). *Electron microscopy and analysis.* Taylor and Francis Ltd, London, UK.

Vickerman, J. C. (1987). Secondary ion mass spectrometry. *Chemistry in Britain*, **10**, 969–974.

Vickerman, J. C. and Briggs, D. (2001). *ToF-SIMS surface analysis by mass spectrometry.* IM Publications, Chichester, UK/Surface Spectra, Manchester, UK.

Walls, J. M. (1989). *Methods of surface analysis.* Cambridge University Press, Cambridge, UK.

Werner, H. W. and Garten, R. P. H. (1984). A comparative study of methods for thin-film and surface analysis. *Rep. Prog. Phys.*, **47**, 221–344.

Surface Analysis on the Internet

The internet provides a fertile source of information on the surface analysis methodologies featured in this book. Sits that are of particular use are those of universities with surface analysis activities, instrument manufacturers and service providers. To list all relevant sites is a vast, and probably impossible, task and to simplify the situation the reader is referred to a single site; that of the UK Surface Analysis Forum at www.uksaf.org. The site is divided into a series of headings covering Techniques, Tutorials, Databases, Software, Journals, Conferences, Academics, What's New and current and past issues of the UKSAF Newsletter. The UKSAF was formed in 2000 by the amalgamation of the ESCA Users Group (founded in 1979) and the UK SIMS Users Forum (established soon after), and provides a focal point for users of XPS, AES and SIMS in the UK although its membership is global. UKSAF holds one day meetings every six months, which include formal talks, workshops and student competitions. Interested parties can join UKSAF via the website and will then be automatically e-mailed details of the biannual meetings.

Documentary Standards in Surface Analysis

The provision of documentary standards at the international level is the responsibility of Technical Committee 201 of the International Standards Organisation (ISO TC201), full details of which can be found at www.iso.ch. From an analytical point of view documentary standards are essential in any laboratory which runs a quality scheme for the following reasons:

1. To improve reliability of the analytical results obtained.

2. To reduce the level of skill required to perform routine analysis.

3. Data can be transferred between different analytical laboratories, with a high degree of confidence if all laboratories follow a similar procedure.

The Scope of TC201

Standardisation in the field of surface chemical analysis in which beams of electrons, ions, neutral atoms or molecules, or photons are incident on the specimen material and scattered or emitted, electrons, ions, neutral atoms or molecules or photons are detected.

The Purpose of TC201

1. To promote the harmonization of requirements concerning instrument specifications, instrument operations, specimen preparation, data acquisition, data processing, qualitative analysis, quantitative analysis, and reporting of results.

2. To establish consistent terminology.

3. To develop recommended procedures and to promote the development of reference materials and reference data to ensure that surface analyses of the required precision and accuracy can be made.

The inaugural meeting of ISO TC201 was held in 1993 and the Committee now comprises of 8 Sub-Committees (SC) and one Working Group, covering all aspects of surface analysis. The two of most relevance to this text are SC5 on AES (chaired by Dr C J Powell of NIST) and SC7 on XPS (Chaired by Professor J F Watts). The Glossary of Chapter 7 of this text was taken from a standard (ISO 18115) prepared by SC1 on Terminology chaired by Dr M P Seah of the National Physical Laboratory of the UK. The current chairman of ISO TC201 is Professor Shimizu of Osaka University and the current Secretary is Mr Yukio Hirose of the Japanese Standards Association, (y_hirose@jsa.or.jp).

Appendices

Appendix 1:   Auger Electron Energies

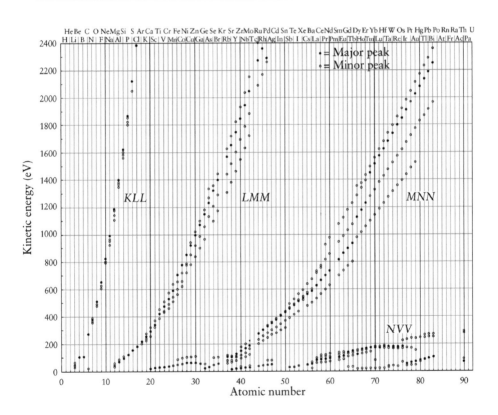

Appendix 2: Table of Binding Energies Accessible with AlKα Radiation

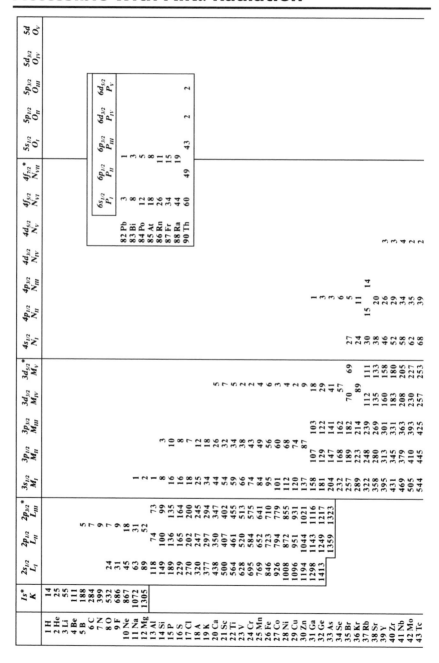

Binding energies (eV) of inner-shell electrons, elements 44 Ru – 90 Th.

Element	$3s_{1/2}$ M_I	$3p_{1/2}$ M_{II}	$3p_{3/2}$ M_{III}	$3d_{3/2}$ M_{IV}	$3d_{5/2}$* M_V	$4s_{1/2}$ N_I	$4p_{1/2}$ N_{II}	$4p_{3/2}$ N_{III}	$4d_{3/2}$ N_{IV}	$4d_{5/2}$ N_V	$4f_{5/2}$ N_{VI}	$4f_{7/2}$* N_{VII}	$5s_{1/2}$ O_I	$5p_{1/2}$ O_{II}	$5p_{3/2}$ O_{III}	$5d_{3/2}$ O_{IV}	$5d$ O_V
44 Ru	585	483	461	284	279	75	43		2								
45 Rh	627	521	496	312	307	81	48		3								
46 Pd	670	559	531	340	335	86	51		1								
47 Ag	717	602	571	373	367	95	62	56	3								
48 Cd	770	651	617	411	404	108	67		9								
49 In	826	702	664	451	443	122	77		16					2			
50 Sn	884	757	715	494	485	137	89		24					1			
51 Sb	944	812	766	537	528	152	99		32					2			
52 Te	1006	870	819	582	572	168	110		40					2			
53 I	1072	931	875	631	620	186	123		50					3			
54 Xe	1145	999	937	685	672	208	147		63					7			
55 Cs	1217	1065	998	740	726	231	172	162	79	77			1	13	12		
56 Ba	1293	1137	1063	796	781	253	192	180	93	90				17	15		
57 La	1362	1205	1124	849	832	271	206	192	99				1	15			
58 Ce	1435	1273	1186	902	884	290	224	208	111				7	20			
59 Pr		1338	1243	951	931	305	237	218	114				12	23			
60 Nd		1403	1298	1000	978	316	244	225	118				14	22			
61 Pm			1357	1052	1027	331	255	237	121				18	22			
62 Sm			1421	1107	1081	347	267	249	130				23	22			
63 Eu				1161	1131	360	284	257	134				33	22			
64 Gd				1218	1186	376	289	271	141		1		38	21			
65 Tb				1276	1242	398	311	286	148		2		38	26			
66 Dy				1332	1295	416	332	293	154		4		38	26			
67 Ho				1391	1351	436	343	306	161		4		39	20			
68 Er				1453	1409	449	366	320	177	168	5		51	29			
69 Tm						472	386	337	180		6		53	32			
70 Yb						487	396	343	197	184	7		53	23			
71 Lu						506	410	359	205	195	7		57	28			
72 Hf						538	437	380	224	214	19	18	65	38	31	5	
73 Ta						566	465	405	242	230	27	25	71	45	37	7	
74 W						595	492	426	259	246	37	34	77	47	37	6	
75 Re						625	518	445	274	260	47	45	83	46	35	6	
76 Os						655	547	469	290	273	52	50	84	58	46	4	
77 Ir						690	577	495	312	295	63	60	96	63	51	0	
78 Pt						724	608	519	331	314	74	70	102	66	51	4	
79 Au						759	644	546	352	334	87	83	108	72	54	2	
80 Hg						800	677	571	379	360	103	99	120	81	58	3	
81 Tl						846	722	609	407	386	122	118	137	100	76	16	13
82 Pb						894	764	645	435	413	143	138	148	105	86	22	20
83 Bi						939	806	679	464	440	163	158	160	117	93	27	25
84 Po						995	851	705	500	473	184	184	177	132	104	31	
85 At						1042	886	740	533	507	210	210	195	148	115	40	
86 Rn						1097	929	768	567	541	238	238	214	164	127	48	
87 Fr						1153	980	810	603	577	268	268	234	182	140	58	
88 Ra						1208	1058	879	636	603	299	299	254	200	153	68	
89 Ac						1269	1080	900	675	639	319	319	272	215	167	80	
90 Th						1330	1168	968	714	677	344	335	290	229	182	95	88

The most intense lines are marked with an asterisk. Thus for elements He to Mg the $1s$ orbital is usually studied, for Al to As the $2p_{3/2}$, and so on

Index